ACCLAIM FOR

ADAM RUTHERFORD

"An enthusiastic guide and a good storyteller."
—*The Wall Street Journal*

"Rutherford unpeels the science with elegance." —*Nature*

"With intellectual authority [. . . Rutherford] expresses himself in a clear and persuasive manner with natural charm." —*The Spectator*

"Rutherford writes with clarity, authority and humor. His research is thorough and so current [. . .] readers will be wowed." —*BookPage*

"Rutherford has a knack for making complex ideas understandable—and also fun."
—Peter Frankopan, Oxford professor of global history

"A master storyteller." —Hannah Fry, author of *Hello World*

"Rutherford is an engaging, witty writer. He is also a concise one." —*The Guardian*

"A superb communicator who eruditely explores the borderlands of history, archaeology, genetics, and anthropology."
—Dan Snow, host of the podcast *Dan Snow's History Hit*

"Rutherford is a delightful, eclectic, hilarious, and often filthy guide to what we know about human genetics."
—Naomi Alderman, author of *The Power*

"Rutherford is a bold, confident storyteller." —*Genome*

Also by Adam Rutherford

The Book of Humans: A Brief History of Culture, Sex, War, and the Evolution of Us

A Brief History of Everyone Who Ever Lived: The Human Story Retold Through Our Genes

Creation: How Science Is Reinventing Life Itself

How to Argue With a Racist

What Our Genes Do (and Don't) Say About Human Difference

Adam Rutherford

THE EXPERIMENT
NEW YORK

How to Argue With a Racist: *What Our Genes Do (and Don't) Say About Human Difference*

Page 209 is a continuation of this copyright page.

Originally published in Great Britain by Weidenfeld & Nicolson, an imprint of the Orion Publishing Group Ltd., a Hachette UK company. First published in North America in revised form by The Experiment, LLC.

The Experiment, LLC
220 East 23rd Street, Suite 600
New York, NY 10010-4658
theexperimentpublishing.com

Library of Congress Cataloging-in-Publication Data available upon request

ISBN 978-1-61519-671-5
Ebook ISBN 978-1-61519-672-2

Cover design by Beth Bugler
Text design by Jack Dunnington
Author photograph by Stefan Jakubowski

Manufactured in the United States of America

First printing April 2020
10 9 8 7 6 5 4 3 2 1

For Ananda, Ben, Jake, Nathaniel, and all of my more distant sisters and brothers

The world is a fine place and worth the fighting for.

–Ernest Hemingway,
For Whom the Bell Tolls

Contents

A Note on Language

This is a deliberately concise book, and I have selected only arguments and cases that are illustrative. I will be using terminology that is not without historical baggage, some of which is considered deeply offensive. I will be using words such as "Black" and "East Asian" while simultaneously acknowledging that they are poor scientific designations for the immense diversity within these billions of people. It is an irony that we roughly know what these descriptors mean colloquially while they are potentially incoherent in terms of scientific taxonomy. The semantics of this book and the broader public discourse are important, and one influences the other. The decision to capitalize Black, White, and any other color named in this book, when it stands for a group of people or a race as it's culturally described, is more than a mere typographical choice; it's a reflection of shifting attitudes regarding the way capitalization—or lack thereof—can either reinforce or undermine a sense of equal respect for all races. And although much of the book discusses the validity of the term "race," I will be using it, primarily because it is a term that people recognize and use, regardless of its scientific validity. "Population," "ancestry," and "lineage" are all terms that are more useful as discussions of human evolution and diversity become more technical. This book is largely focused on racism derived from Western and European cultures, partly because these are my cultures, but also because the concepts of race that we are broadly globally wedded to emerged in Europe and were enshrined in culture alongside European expansion, the emergence of science as we recognize it today, and the values of the Enlightenment.

Preface

I write these words days before the summer solstice of 2020. Though the year is only half done, the world has already been rocked twice by events with race at their core: a pandemic, which has threatened every human alive but has killed with discrimination, and protests against police brutality that raged after a White cop pressed his knee down on a Black man's neck for eight minutes and forty-six seconds, killing him. Frustration, anguish, and anger are appropriate responses to these two situations. And yet, as shocking as they are, the racism involved is nothing new; issues about race, racism, ancestry, and genetics have become more and more prominent in the public consciousness in the last few years—a disturbing trend that compelled me to write this book. I want to show that although, historically, science has been misused to institutionalize racism, today, science is no ally to racists. In my view, science can and should be deployed as an anti-racist tool.

In January, the gears of the world began shutting down. Many scientists—and some politicians—already knew that another pandemic was imminent and inevitable, though few predicted the impact that COVID-19 would have on all our lives. As I write, we are a long way from seeing how this will unfold: when or whether we will create a vaccine, when or whether there will be a second wave—or multiple waves—or if this disease will become a permanent specter in our lives. Debates rage about the science and the policies that could, should, and have been implemented, and two countries among those hit hardest—the US and the UK—account for more than a third of the global death toll. As of writing, more than 7 million people in 188 countries have been infected, of whom more than 400,000 have died.

And then, at the end of May—while governments scrabbled around with varying degrees of efficacy—a police officer in Minneapolis revealed once more how deadly the combination of racism and power can be, when he crushed the life out of forty-six-year-old George Floyd. All it took was the officer's own silence and weight. Demonstrations in America and around the world erupted, the uprisings show no signs of abating, and race dominates our public discourse again. I'll return to these urgent and vital protests—and the racist pseudoscience helping fuel the dangerous way police treat Black people as "other"—in the pages that follow.

The disease COVID-19 and the coronavirus that causes it were first identified in the city of Wuhan in China in December 2019, and it rapidly became racialized via two discrete routes. The first was that the provenance of the virus became the source of hostility in casual and, in some cases, extreme ways. The science is far from clear about precisely where this new coronavirus came from, but bats are a likely reservoir; at this time, our best guesses are that it leapt the species barrier from bats to humans via the Huanan Seafood Wholesale Market, a "wet market" in Wuhan, where meat and seafood are traded. It may have been transferred via scaly mammals called pangolins, though none are listed in the market's inventory (perhaps omitted because pangolin trading occurs illegally, albeit persistently). As the disease spread and anxiety levels increased, there were calls from the West to ban wet markets—which mostly failed to recognize that, in general, this term is used to distinguish trading posts that sell fish and meat from those with electrical goods, clothes, and so on, as well as from supermarkets selling dry or frozen food. Though the Huanan market also sold wild animals—which may yet prove to be the source of the spillover from animals to humans—the West's misunderstanding of wet markets became a pivotal point for the cause of racial antagonism.

Journalists suggested that President Trump (and other public figures) exacerbated xenophobia by referring to the novel coronavirus as "the Chinese virus," potentially putting Asian Americans at risk of targeted attacks. Trump defended his choice of words: "It's not racist at all. . . . It comes from China, that's why." Others supported Trump's position by pointing out that one of the deadliest pandemics of the modern era is referred to as "the Spanish flu." As an alibi, this holds no water. The name came into use not because the strain originated there but because Spain retained a free press during the First World War and reported the flu openly, when censorship was imposed elsewhere. The provenance of that influenza virus remains unknown, but plausible candidates include France and a military base in Kansas.

The geographical origin of COVID-19 rapidly became an enabling factor for coronavirus-related racist attacks, which are now so numerous that they have their own Wikipedia page. In February, on the street where I work in London, an Anglo-Chinese student from my own university was violently beaten, the four assailants shouting, "I don't want your coronavirus in my country." In the US, Russell Jeung, professor of Asian American studies at San Francisco State University, compiled data revealing that, in the spring of 2020, thousands of racial attacks were indiscriminately directed against Korean Americans as well as Chinese Americans. The FBI discovered evidence that right-wing extremists were calling for attacks on Jewish and Asian Americans—and for the virus to be intentionally spread in synagogues and mosques.

The second way that the novel coronavirus became racialized was less bound to hatred and more derived from peculiarities in those who were infected. As the disease spread, the emerging picture suggested that Hispanic or Latino, Black, Asian, and other people of minority ethnicities were significantly more at risk than those with primarily White European ancestry. From the beginning of the pandemic, as early as April, striking disparities were seen: In Chicago, where one third of the population is Black, almost three quarters of people who died were Black. In New York, almost twice as many hospital admissions were Hispanic or Latino compared to White people. These types of statistics have replicated to varying degrees all around the world.

As these discrepancies came into focus, some people took them as a definitive signal that race is indeed a biologically informed category—contrary to the clear evidence of contemporary human genetics. This book is about race and its longstanding, knotty relationship with basic biology, evolution, and genetics. It is a book about how the history of race science has sought a biological foundation to legitimize the racial categories that humans invented—and how genetics can be co-opted, distorted, and misrepresented to fulfill that quest. When properly understood, modern genetics refutes any meaningful biological basis for racial categories.

The fact of different infection and death rates in minority groups is important and interesting, but using data from COVID-19 to crank out obsolete and fractally incorrect assertions about race is absurd. Even people who tenaciously cling to races as biological categories do not lump Black, Asian, and Hispanic or Latino people into one group. To suggest that apparent increased vulnerability to the novel coronavirus is evidence for biological races would serve only one purpose: to separate White people from everyone else.

One idea that has been proposed to explain this question of ethnicity and COVID-19 involves vitamin D, which has known antiviral properties. We know that vitamin D production is stimulated by ultraviolet light from the sun—and that melanin inhibits its production, such that people with darker skin sometimes have vitamin D deficiency. This theory is worth considering, but *if* it turns out to be valid, it does not racialize COVID-19. Rather, it simply provides a biological basis for a slight increased risk in all people: Vitamin D deficiency also more significantly affects men than women, people with obesity and type 2 diabetes, and other categories of people who seem to have elevated COVID-19 risk. In any case, *if* that theory turns out to be valid, it will account for only a small portion of the disparities we are seeing.

In contrast, we *know* that well-established social and cultural phenomena have very significant negative effects on health in minority communities. People from these groups are much more likely to be essential workers and therefore, they were not under enforced lockdown. In addition, social isolation hasn't been an option for them to the same extent that it has been for those of higher socioeconomic status. Minority groups tend to live in densely populated urban areas—often in effectively segregated housing—where social distancing is more difficult to practice. They are more likely to live in multigenerational households, again making social distancing more difficult and increasing the risk to elderly people. Alongside poverty and other social phenomena, these factors are well-known to negatively correlate with health and life expectancy. None are unique to COVID-19. A preliminary study in the UK showed that the increased risk of death among Black people vanished if social deprivation and other underlying health conditions were taken into account.

These are very early days in our understanding of such a serious pandemic. At the moment, the best we can say with certainty is that the underlying causes for the disproportionate effect of the novel coronavirus on non-White patients are many. Genetics—possibly in the form of vitamin D metabolism—may play a small part in that mix, which *may* relate to pigmentation in some small way alongside myriad hugely significant social factors. But we can also say with certainty that this disease does not demonstrate a biological basis for traditional racial categories. No diseases do so, as discussed throughout this book. The full picture is yet to be revealed, and it will be many years before we understand this devastating pandemic. Simplistic, racialized explanations offer little value. As Charles Darwin wrote 150 years ago, "Ignorance more frequently begets confidence than does knowledge," and this remains as applicable today as ever. The truth is that the underlying reasons fully explaining the disproportionate impact of COVID-19 on certain populations can be summarized in the three most important words a scientist can say: We don't know.

AR
June 2020

Introduction

This book is a weapon. It is written to equip you with the scientific tools necessary to tackle questions on race, genes, and ancestry. It is a tool kit to help separate fact from myth in understanding how we are similar and how we are different.

Our story began in Africa. The earliest known members of our species—*Homo sapiens*—evolved in what is now Morocco around three hundred thousand years ago, though most early remains are from the east of Africa. We are starting to think that in the beginning, we came from a pan-African species, a mixture of diverse populations from around that mighty continent. We know that some early humans migrated into Asia and Europe within the last quarter of a million years, but their dominion outside of Africa was temporary, and they probably leave no descendants today. Around seventy thousand years ago another group of people drifted away from Africa, and the process of setting down new roots all over this planet began. Much of our global success is a result of local adaptations, fine-tuned by evolution to best survive environments on an ecologically diverse planet. Our quintessential nature as wanderers, hunters, farmers, and social creatures means that, over the last few thousand years, Earth has become smaller, and peoples from around the world have met, traded, mated, fought, conquered, and a whole lot more. In these interactions, we engage with people who are different from each other. These differences are rooted in biology, in DNA, and also in our behavior as social animals—in our dress, our speech, our religions, and our interests. In the pursuit of power and wealth, the fetishization of these differences has been the source of the cruelest acts in our short history.

The political climate has changed in the past few years. Around the world, nationalism is on the rise, and the presence of race in the public arena is more prominent than it has been in some time. As I write these words, cities around America are electrified by popular uprisings. Hundreds of thousands have marched, protested, and rioted (though most are impassioned but peaceful demonstrations)—this time, prompted by the killing of George Floyd by a police officer in Minneapolis. This is a race conflict—as it was in 2016 in Charlotte, North Carolina, after police killed an unarmed Black man named Keith Lamont Scott. Or as occurred in Ferguson,

Missouri, in 2014, after unarmed teenager Michael Brown was killed by police, just as had happened to Freddie Gray in Baltimore in 2015 and Timothy Thomas in Cincinnati in 2001. Anti-racist protests sprung up all around the US in 1992, after the brutal assault of Rodney King by four Los Angeles police officers—and in Miami in 1980, after four cops beat Arthur McDuffie to death for running a red light. For all these incidents, the police were either acquitted or not criminally charged. At present, the fate of the four officers involved in the killing of George Floyd remains to be seen.

These are race riots—as they arose all around the country in 1968, following the murder of Martin Luther King, Jr.

In one sense, nothing has changed. The US has never resolved its racist history, and the daily frustration of normalized prejudice endured by Black people—and other non-White Americans—tipped over into public protest and violence in May 2020, as it has many times before. Today, though, perhaps unlike the racial violence of the twentieth century, factions enabled by technology are born and nurtured on social media. Black Lives Matter began as a hashtag, following the acquittal in 2013 of the man who fatally gunned down Trayvon Martin a year and a half earlier, and it has escalated into a global movement—its stated aim: to vanquish White supremacy and counter acts of violence against Black people.

Rodney King's beating was taped—on a shaky 8-millimeter camera at a distance—and this was a foreshadowing of what would come. George Floyd's death was filmed on multiple cameras: a police officer's knee crushing his throat for almost nine minutes. The footage was broadcast globally within hours, and his words—"I can't breathe"—echoed Eric Garner's exact words before his death from a police officer's chokehold, on Staten Island in 2014, and thus, the slogan was revived for demonstrators around the world. The actions of protestors and police are now continually documented in a chaotic melee for all to see—the protests, the looting, the police brutality. The schisms of a country built on racist foundations—on the backs of the enslaved—are now more than ever exposed, like a nerve. The revolution was not televised; it was livestreamed.

The protests represent the outcome of systemic, structural racism in our societies, which is not simply born of acts of violence against Black men by the police—nor from the voices of White supremacists that become mainstream. Structural racism is everyday—and rooted in the everyday. It is rooted in indifference to the lived experience of the recipients of racism.

Stereotypes and myths about race are the foundations on which structural racism is built, and these have been ingrained in western culture, while laced with centuries of pseudoscience that I will dissect in these pages. We see the raw exposure of racism in both these acts of police violence and the subsequent protests and riots, but the misguided and malicious views from which these events emerge are pervasive and stubbornly entrenched. The insistence that outdated racial categories are rooted in biology is upheld not just by overt racists whose voices are amplified by modern technology but also by well-intentioned people whose experience and cultural history steer them toward views that are simply not supported by the contemporary study of human genetics: the misattribution of athletic success to ancestry rather than training, or the continued assumptions that East Asian students are inherently better at math, or the idea that Black people have some kind of natural rhythm, or the notion that Jews are good with money. We all know someone who thinks along these lines. The ideas examined in these pages form a scientific description of real human similarities and differences that will provide a foundation to contest racism that appears to be grounded in science. Here, I am focusing on four key areas where we often slip up by adhering to stereotypes and assumptions; I am outlining what we can and cannot know according to contemporary science on the subjects of skin color, ancestral purity, sports, and intelligence.

It is often easier to make a claim than to refute it, but as racism is being expressed in public more openly today, it is our duty to contest it with facts and nuance, especially if bigotry claims science as its ally. Some scientists are not comfortable with expressing opinions derived from their research where it relates to questions of race. Nevertheless, if you study human genetics—the ocean from which human variation is drawn—you have little choice but to speak of race.

The visible differences that are the roots of racism are encoded in our DNA. Therefore, science and racism are inherently entwined. Racism is an expression of prejudice, whereas science, in principle, is free from subjectivity and judgment. Reluctance by scientists to express views concerning the politics that might emerge from human genetics is a position perhaps worth reconsidering, as people who misuse science for ideological ends have no such compunction and embrace modern technology to spread their messages far and wide.

But science is a powerful ally, and knowledge of science and of history arms us against preconceptions and prejudice. We have profoundly limited senses and short lives. We crave meaning, and belonging, and identity. Those aspects of the human condition are a rich soil in which prejudice can take root. The tool that grants us the clearest view of how people actually are, rather than how we judge them to be, is science.

I am British. My identity is legally enshrined in my passport, the property of the United Kingdom. It was issued in Ipswich, a town near the East Anglian coast, where I was born.

These are facts. Britain, United Kingdom, Ipswich, East Anglia—they are labels that partly define my personal identity. I am also a scientist. I have studied genetics and evolution all my adult life, and I write about how history intersects with those two forces of biology.

In science, we use labels out of necessity. We try to apply rigorous criteria in our labeling to help us categorize the inherent qualities of a thing, so that we might understand its identity, its essential nature, or its evolution, or so that we can design experiments that will help us understand its qualities. We call this "taxonomy."

I am mixed race, or dual heritage, or biracial. "Half-caste" is a term that has fallen out of favor, but for much of my life that is how many have described me, some out of habit, occasionally in a dismissive way. I am often asked where I am from, and I adjust my answer by second-guessing what they are really asking: Britain, England, Suffolk, Ipswich, or London, where I have lived for twenty-five years. All are true, but often, what they are really asking is: Why do you look the way you do? My father was born in Yorkshire, in northern England, with both his parents being White and British. My mother is British and Indian, though she has never set foot in India. She was born in Guyana in South America. Her grandparents were shipped there from India in the nineteenth century to work on sugar plantations under a colonial edict known as indenture—a form of semi-forced migration and labor that is a shadow of slavery. She emigrated to England in the 1960s, in the wake of the *Empire Windrush*, the ship that brought 802 Caribbean women and men to begin new lives in Britain in the aftermath of the Second World War. Like them, she was a British citizen invited to the homeland of the colonies as the imperial age waned. They were bidden to help rebuild a country broken by war, and like so many who made that journey, my mother was recruited into the fledgling National Health Service, to attend to the citizens of the United Kingdom.

My parents did not stay together, and when I was young, my father, sister, and I merged into a new family, where I acquired three more brothers (though technically they are two stepbrothers and a half brother). I lived in Ipswich until I was eighteen, and then went on to study genetics at University College London. I have remained in London and tethered to UCL ever since. I do not consider myself to be half-caste, or half anything. The nature of my upbringing has been that I have no cultural affinity with India, though, like so many Brits, I love two things that India does better than any other nation—curry and cricket. Yet it is undeniable that biologically, half of my DNA is more closely associated with 1.3 billion Indian people than it is with 740 million Europeans—and of course the obverse is also true.

I did not study genetics and evolution because of my heritage; I chose it because it is by far the most interesting branch of scientific research, one that underwrites every single aspect of the life sciences. “Nothing in biology makes sense except in the light of evolution,” said the Russian American scientist Theodosius Dobzhansky, a mantra that should be as widely known as any. I got lucky in that I stumbled into a field of science at a time when it was about to enter a golden age of discovery. The Human Genome Project officially started the year I went to university, and its fruit—a draft of the entire genetic code of a human being—was completed the year I finished my PhD, also in human genetics. I used that DNA database extensively to hunt for genes that build our eyes and govern how we see. Since then, that colossal, glorious scientific endeavor and the technology and data that followed have formed the bedrock on which future genetics and in turn all biology would forever be transformed.

UCL is one of the great universities. The foundations of genetics and evolution were developed there during the first half of the twentieth century, as Darwinian ideas were fused with the emerging concept of genes, via statistics, experimentation, and math. There, on Gower Street in the West End district of Bloomsbury, much of the structure of modern biology was being cooked up.

But some of the most pernicious ideas in human history also have deep roots at UCL; most significantly, it was intrinsically associated with the birth of eugenics, the idea that, via selective breeding, human populations could be improved and weakness eliminated from societies. These ideas were primarily formulated in Britain, by the scientist and avowed racist Francis Galton, though they were never enshrined in law there. The UK came perilously close in 1912: The Mental Deficiency Bill was brought before Parliament, with a eugenics amendment that would prohibit marriage and procreation between the "feebleminded," as was the vernacular of the time. The clause was removed by the member of parliament Josiah Wedgwood before the bill was passed into law in 1913. In contrast, the governments of the US, Sweden, Nazi Germany, and other countries had active eugenics policies that resulted in the forced sterilizations and deaths of millions. Though these policies are regarded as pernicious today, forced sterilization continued throughout most of the twentieth century in the US. Thousands of American Indian women underwent involuntary sterilization up until the 1970s, as did African American women with multiple children, under the threat of withheld welfare. In California, an African American woman was sterilized without consent in 2010. Eugenics and racism are not the same ideas, but they are inherently connected, and eugenics policies disproportionately affected and targeted racial minorities.

I did not choose UCL because of its peculiar history, though I was enrolled in the Galton Laboratory, which was once called the Galton Eugenics Laboratory, and was taught by the Galton Professor in the Galton

Lecture Theatre, all named after Francis Galton—a man whose intellectual legacy includes weather maps, a phalanx of essential statistical techniques, forensic fingerprints, and the scientific concept of eugenics, as well as the word itself. Galton died in 1911, and the men who followed him at my alma mater were similarly great scientists: the statistician Karl Pearson, the mathematical biologist Ronald Fisher, and others, men on whose shoulders stand entire domains of contemporary science, and who, to varying degrees, also expressed racist views. To call them racist is not a judgment based on contemporary sensibilities, it is a factual statement; they articulated opinions that were racist, as were the cultural and scientific norms at the time.* Science is

*There are many examples of Galton's racist views, but they are perhaps most explicitly stated in his letter to *The Times*, June 5, 1873, headlined "Africa for the Chinese": "My proposal is to make the encouragement of the Chinese settlements at one or more suitable places on the East Coast of Africa a part of our national policy, in the belief that the Chinese immigrants would not only maintain their position, but that they would multiply and their descendants supplant the inferior Negro race. I should expect the large part of the African seaboard, now sparsely occupied by lazy, palavering savages living under the nominal sovereignty of the Zanzibar, or Portugal, might in a few years be tenanted by industrious, order loving Chinese . . . Of all known varieties of mankind there is none so appropriate as the Chinaman to become the future occupant of the enormous regions which lie between the tropics . . . The Hindoo cannot fulfil the required conditions nearly as well as the Chinaman, for he is inferior to him in strength, industry, aptitude for saving, business habits, and prolific power. The Arab is little more than an eater up of other men's produce; he is a destroyer rather than a creator, and he is unprolific."

wont to change as new data becomes available. By the 1990s, we studied these men's scientific legacies while acknowledging that their attitudes and beliefs were racist. Their views were not shared even superficially by any of the scientists who taught me.

That I am both British and a geneticist are two objective facts laden with centuries of context. I am the evolutionary descendant of colonialism, empire, racism, and some pretty odious ideologies. My own story is not particularly unusual or interesting—politics and families are messy, people move, fall in love, have children, and repeat any or all of these within or between generations. That is all the biographical information needed for this story, but in some senses, all my multiple lines of ancestry—biological, cultural, and scientific—have inevitably clashed. I haven't endured a great deal of racial abuse in my life—I am light-skinned and my Indian (or Indo-Guyanese) heritage is far from obvious. But in the last couple of years, in response to my writing and talking about human history, genetics, and race, strangers have called me a Paki, a Jewish rat, and a race traitor with "insidious influence." My Indian heritage is not Pakistani as far as we know, I have no significant Jewish ancestry (though my stepfamily does), and I believe my alleged racial treachery is that I married a White Englishwoman. I have been told that I should be grateful for colonialization and the British Empire as, without it,

I would not exist—technically this argument is correct, though it's pretty nuts.

The cultural conversation has changed in recent years, and the vocal expression of racism feels more prevalent today than it has been in decades. In 1939, Agatha Christie published her bestselling thriller *Ten Little Niggers*, which remained in print in the UK with that title until 1963, before becoming either *Ten Little Indians* or *And Then There Were None*, which was the primary title in the US. A year later, a British Conservative member of Parliament was elected to represent the Smethwick area of Birmingham with campaign leaflets bearing the slogan "If you want a nigger for a neighbour, vote Labour" (the main liberal party).

The racial histories of the US and Britain are very different, but there are common themes. Britain never had formal racial segregation, and no specific civil rights movement, which in America helped bring about the landmark Civil Rights Acts of 1964 and 1968, prohibiting many major forms of legal segregation. Sports remain a common and high-profile battleground for racial issues; in baseball, there was the unofficial "color line"—a tacit, unwritten agreement that segregated Black players from White (and American Indian and Hispanic or Latino players, such as Lefty Gomez). That ended in 1945 when Jackie Robinson signed for the Brooklyn Dodgers. We don't play baseball in the

UK, but I grew up watching soccer in the 1980s, in stadiums where thousands of fans would shout "Shoot that nigger" about Black players on their own team. Race hatred trumped team loyalty. At my school, some boys would play a game where they would leave a two-pence piece on the ground, and then shout "Jew" or "kike" at anyone who unwittingly picked it up.

We like to think explicit racism is no longer openly part of culture, society, or sport, though in 2018 banana peels were again being thrown onto soccer fields at Black players in the UK, as they routinely were three or four decades ago, to assert that the players are closer to monkeys than humans.* Similarly, in US schools in 2019, there were reports of bunches of bananas being left in locker rooms and watermelons being brandished when White football teams played against predominantly Black ones. And in basketball, a Cincinnati school team was kicked out of a youth league for fielding players with jerseys featuring nicknames including "coon" and "knee grow."

It's not easy to assess how racist a society is; people are reluctant to volunteer information (even anonymously) that might be perceived as culturally unacceptable. Nevertheless, Pew surveys in the US in 2019 indicate that

---

*This is a terribly misguided attempt at bigotry: Edible bananas are an entirely human invention, and monkeys aren't that into bananas anyway. When a banana was thrown at FC Barcelona player Dani Alves in 2014, he had the good grace and composure to eat it.

more than half of Americans think that race relations in the US are currently bad, and have soured during the current administration, and two thirds think that the vocal expression of racist sentiments is now more common in the Trump era. The effect of the demonstrations and riots that followed the homicide of George Floyd in 2020 is yet to be fathomed—but again reveals deep schisms in American society.

University of Illinois surveys have tracked some attitudes to race since the 1940s, and the trajectory is unequivocally toward racial progress: In 1942, only a third of White Americans thought that Blacks and Whites should go to the same schools, but by 1995 (the most recent time the question was asked), the number was 96 percent. In 1954, less than one in twenty White Americans approved of interracial marriage; in the last decade the proportion is more like eighteen out of twenty. In similar British national attitude surveys that have been running since 1983, the proportion of people who describe themselves as either "not prejudiced against people of another race at all" or "very or a little racially prejudiced" has remained static (60–70 percent and 25–40 percent, respectively). Instead, we might use proxies such as asking people whether they would be happy if a close relative were to marry a person of Black or Asian background. In 2017, over a fifth of White Brits answered that they would mind. This is a racist view, but when that question was asked in the equivalent survey in 1983, the answer was more than 50 percent. The same question was asked in 2017 (but not before then) about the prospect of a Muslim spouse, and the answer was more than two fifths of people would be bothered.*

These are complex metrics that broadly indicate attitudes toward race are relaxing in some directions, and they reflect the fact that culture changes. Another somewhat informative statistic in the UK is that the British police indicated that reports of racist attacks went up in 2016 around the time of the referendum on Britain's membership of the European Union, though this cannot be definitively interpreted as an increase in racism in the UK—it could be that the frequency of crime is static, but willingness to report it has increased, emboldened by positive responses from the police.

The nineteenth-century abolitionist preacher Theodore Parker said that the moral arc of the world tends toward justice, and while this may be true, it does not mean that bigotry evaporates. It merely re-arms according to

---

*It is sometimes argued that as Islam is a religion, then Muslims are not a race; therefore prejudice against Muslims cannot be racism. This is true in a very literal sense. However, for the purposes of the comparison between the 1983 and the 2017 statistics, the bigotry and prejudice against a specific group of people often resembles racism very closely, even though they might not be categorized as a traditionally defined race. Similarly, prejudice against the people of a specific country—for example, Romanians—is effectively an equivalent form of cultural racism.

the prevalent culture—White people in Britain are apparently less comfortable today with Muslims than they are with Black or Asian Brits who are not Muslim. Concepts of race have always been associated with attempts to categorize humans, sometimes to simply describe them, often to create pseudoscientific delineations, with the intention of subjugation and exploitation.

While it may be difficult to assess how racist a people is, and whether that is changing, we *can* track with perfect accuracy how science changes. Discoveries are made, knowledge is created, techniques evolve, and all of it is documented meticulously. The field of genetics, with its racist past, has undergone radical transformation in its short history. It has not only grown to be intrinsic to scientific research; it is also now integrated into the wider culture and has become a huge commercial business endeavor targeted at ordinary people. We know more about human variation, migration, and history than ever before, and this exposure has invigorated questions of race.

Genetics is merely the scientific study of families, sex, and inheritance, all ideas that preoccupied human minds for millennia before Darwin, Mendel, Watson and Crick, and the other scientific pioneers that ushered in the current era. Human genetics is the study of similarities and differences between people and populations. There were major transitions in genetics in the twentieth century: the discovery of the structure of DNA, the cracking of the genetic code, the birth of the endeavor to read the entirety of human DNA. These were necessary preludes to a perpetual revolution occurring in genetics in the twenty-first century. Following the Human Genome Project, our ability to sequence and understand DNA has exploded beyond any expectations that we might have had in the 1990s. We have the genetic code of millions of people in databases that scientists pore over and mine for wispy clues about diseases, behavior, and ancestry. Even more unanticipated, a growing number of these people are dead, and have been for hundreds, thousands, or even tens of thousands of years. The DNA from those old bones provides incomparable data on our history and prehistory, on how we migrated from Africa and flooded the earth. These records tell us what people were like before we began documenting our lives.

Most scientific research is done in the public domain, and most genomic databases are open for all to mine. But they are dwarfed by the number of genomes that have been sampled and are owned by a handful of genetic genealogy companies, who, for around a hundred bucks and a tube full of spit, will provide you with a map of the people on Earth whose DNA yours most closely resembles. Direct-to-consumer genetic testing is a murky area scientifically and ethically, prone to gross simplification and romantic storytelling, and I will scrutinize it later in this book.

Millions of people have paid for and taken these tests. I spend a lot of time talking about genetics in public, and I've observed an interesting phenomenon. Once you arm people with their own genetic code, which has been inaccessible and incomprehensible until now, the cultural conversations about race, identity, ethnicity, and genetics change. Mostly it's trivial: White people always want to discover that they are descended from Vikings, because let's face it, Vikings were really cool. In part 2, I will explain why *all* people of European descent have Viking ancestors. Irish, Welsh, and Scots like to claim Celtic genetic genealogy, even though "Celtic" isn't a coherent ancestral population; despite cultural similarities, the latest genetic data show that those three groupings are frequently more similar to mainland English people than they are to each other. In this sense, using contemporary genetics to assert these types of cultural identities is not very fruitful, but it is also of little consequence—we desire membership to clans, tribes, and families, and while these narratives can be drawn from geography, nationhood, and history, ancestral genetics says very little about them.

At the far end of the same spectrum, White nationalists and neo-Nazis are also co-opting genetics as a means of asserting their ethnicity, and therefore their supposed racial superiority. In 2018, neo-Nazis in America introduced a new way of showing this off:

They filmed themselves "chugging milk"—that is, gulping down cows' milk with their shirts off in a ridiculous attempt to demonstrate their genetically encoded capacity to process lactose, a sugar in milk that cannot be digested by the majority of humans after weaning, apart from Europeans. The gene mutations that allow this enzymatic ability—known as lactase persistence—arose in Europe around eight thousand years ago, and the ostentatious showcasing of a random mutation that nature selected to allow some people to drink milk throughout life without minor tummy troubles is somehow associated with their assertion of racial superiority. They are presumably unaware that the same mutations emerged independently and exist at a high frequency in Kazakhs, Ethiopians, Tutsi, Khoisan, and populations in many other places where dairy farming was a significant part of their agricultural evolution, including not just milk from cows and goats but camel milk for pastoralists in the Middle East.

Risible though milk chugging is, avowed racists have shown a great interest in modern genetics as a tool in their armory, with a similar degree of misunderstanding of the complexities of human evolution and history as those who simply yearn to be a bit Viking. More broadly, population genetics is being co-opted to reaffirm old and natural tendencies that we have to seek meaning and identity in our societies. Attempts to justify racism have always been rooted in science—or more specifically in misunderstood, misrepresented, or just plain specious science. It never went away, but now we stand at the beginning of the third decade of the twenty-first century, racism is making an overt comeback, revitalized by the new genetics.

This science is hard. It relies on wading through the largest and most complex dataset that we are aware of—the human genome. The tools we apply to extract meaningful information from a code made up of three billion letters are immensely complicated, too, statistical nightmares that require both expertise and deep thought. The history of race, of colonialization, empire, invasion, and slavery is similarly tortuous, and the subject of serious academic scrutiny. But the expression of these disciplines is in everyone's lives. Humans come packed with prejudices, taught, learned, and acquired through experience, and these can form the foundations of views that are not supported by contemporary science.

We crave simple stories to make sense of our identities. This desire is at odds with the reality of human variation, evolution, and history, which are messy and extremely complicated. But they are recorded in our genes. The aim of this book is to anatomize and lay out precisely what our DNA can and can't tell us about the concept of race.

Human genetics is the study of how we are different and how we are the same as each other: in individuals, in disease, in populations, and in history. Most (though not absolutely all) contemporary geneticists disagree with the idea that genetic variations between traditional racial groupings of people are meaningful in terms of behavior or innate abilities. Yet academic papers continue to be published in which genetic bases for complex traits appear to be stratified by racial lines. Though the publication of papers in reputable journals via the process of peer review is the standard way of disseminating research, this is not a marker of some gold standard of truth. Instead, it is a signifier that the research is of a standard worthy of further academic discussion. Genetics is technical and statistical, and there are many ways to cut a cake,* skin a cat, or process a genome-wide association study. Scientists disagree all the time about the significance of results, or the techniques deployed in their analyses. It is perfectly possible for a paper in a reputable journal to be flawed, or even wrong. That

---

*A novel way to cut a round cake was invented and published in the prestigious journal *Nature* in 1906, authored by none other than the father of eugenics Francis Galton, who noted that the "ordinary method of cutting out a wedge is very faulty." It involved cutting a slice from the diameter and squishing the two remaining sections together to retain moisture for the next day. Then do the same but in an axis at right angles to the previous day's slice. Galton's intellectual legacy is both profoundly positive and negative, but alas this cake-slicing technique is not part of it.

is why we publish—so that other experts can test our ideas. As distribution of research is pleasingly easier in the age of the Internet, so also is the dissemination of poor arguments and misinterpretation by bad actors. As a result, the nuances of such academic discussions are lost in a mire of angry, scientifically illiterate assertions of tribalism, identity politics, and pure racism disguised as science.

Often, these discussions are hampered not just by inexpertise but by the imprecision of language. "Race" is a very poorly defined term. Since the seventeenth century, attempts to categorize people into racial types have resulted in the number of races being anywhere between one and sixty-three. We talk casually of Black people, or East Asians, or other categorizations of billions of people that primarily refer either to geographical landmasses or a handful of physical characteristics—none more so than pigmentation.

Racism has many definitions; a simple version is that racism is a prejudice concerning ancestral descent that can result in discriminatory action. It is the coupling of a prejudice against biological traits that are inalterable with unfair behavior predicated on those judgments, and can operate at a personal, institutional, or structural level. By this definition, racism is something that has always existed, even though race as a concept has changed over time. The term "race" has historically been synonymous with more scientific categories such as subspecies or biological type, but these categories have also been used to describe animals and vegetables, as well as tribes, nationalities, ethnicities, and populations.

In modern biology, race has been used with more specificity, as informal categories that people generally understand owing to contemporary common usage. But as a result of ever more precise taxonomy in humans, none of the historical or colloquial usages of race tally with what genetics tells us about human variation. As a result, we are prone to saying glib things such as "race doesn't exist," or "race is just a social construct."

While these sentiments may be well-intentioned, they can have the effect of undermining the scientifically more accurate way of expressing the complexities of human variation, and our clumsy attempts to classify ourselves or others. Race most certainly does exist *because* it is a social construct. What we must answer is the question of whether there is a basis to race that is meaningful in terms of fundamental biology and behavior. Are there essential biological (that is, genetic) differences between populations that account for socially important similarities or divisions within or between those populations?

If race is a social construct, there is a biological basis to that, too: The crude categorization of peoples is done by physical traits such as pigmentation or physiognomy, and we have to acknowledge that these are characteristics that are determined in large part by the expression of genes, which vary between people and populations in ways that we can scrutinize with more depth and accuracy now than at any time in history. Cultural categorizations are mostly derived from ancestry, and this means broadly that people within one group are more similar to each other genetically than they are to people not in that group. Are these variations biologically significant? The dark skin that we most often associate with people whose ancestry is largely not from the Out of Africa diaspora some seventy thousand years ago is determined by genes, as is similarly dark pigmentation in people of south India and the indigenous people of Australia, both of whose ancestors left Africa millennia earlier. No one really thinks that the versions of those pigmentation genes in African people confer the ability to run faster or longer than others. Yet a common assumption persists that there is something implicitly associated with pigmentation that translates to physical abilities. Many influential voices from European history—Kant, Voltaire, Linnaeus—believed this.

We are a rich symphony of nature and nurture—of DNA and environment—stuff we are born with and stuff that happens within us and to us. Our fundamental biology is encoded in our genes, which are inherited from our parents—and therefore ancestors—in a combination that is unique to each one of us. That code is inalterable (short of mutations that can be innocuous or might cause disease, such as cancers), and therefore forms the foundations of our lived lives. There is no perfect metaphor that usefully describes the breathtaking complexity of our genomes, as revealed by twenty-first-century science. People have spoken of DNA as a "blueprint" for years, but this is misleading and has little explicatory value, as it implies a detailed, mapped-out plan, each instruction describing a component of our biology that is determined by its nature.

Genes are sequences of coded chemicals that determine the order of amino acids that form the proteins that enact our biology. The steps from the raw written code to a lived life are extraordinarily complex. Proteins come in the form of enzymes, hormones, cellular architecture, molecular machines, transporters, and all operate in networks with other molecules, in a range of cells and organelles, in tissues and organs, expressed in time and space from conception to death. When we speak of nature and nurture, it is not useful or accurate to think of these two phenomena in opposition. Nature—meaning DNA—has never been *versus* nurture (meaning everything that isn't DNA). Our genomes are the totality of our DNA, and that is where our genes are. Nurture—meaning the nongenetic environment—does not mean whether your parents cuddled you or ignored you as a child; it means every interaction between the universe and your cells, including how you were raised, but also everything from the orientation of you as *fetus in utero* to the randomness of happenstance, chance, and noise in a very messy system.

In the twentieth century, scientists swung between poles of genetic determinism and genetic denialism. The popular eugenics movements in the prewar years typified a belief that our successes and foibles were inbuilt and unchangeable. After the atrocities of the Second World War were exposed, research culture swung toward the "blank slate"—the idea that it is the environment that shapes our character. The truth is, inevitably, somewhere in the middle, though there are ongoing debates about which is dominant. Certainly, to deny the importance of genetics in influencing our behaviors is folly. This is perhaps most obvious in sport, an arena in which the playing field is never level. Success in sport undoubtedly has a fundamentally biological basis—physiology and anatomy are intrinsic to victory. Physical forms vary in populations around the world, and our love of sport can lead us to conjure links between genetics, ancestry, and anatomy. But genes are not the only factor in determining sporting success, which is a profoundly complex interaction between genetics and a life, as we shall explore in part 3. The questions we have to answer in relation to biology, culture, and race concern the weight of the influence of genes, and whether it is unique or essential to certain populations.

If studying humans is complex, there is no area within biology more difficult to understand than our cognitive abilities. The science of understanding how brains work is in its unruly infancy. How neurons connect and harbor thoughts, how those thoughts translate into action or experience within people and between people remains mysterious. Neuroscience, psychology, sociology, and anthropology are all scientific disciplines about people that predate, yet are rooted in, genetics. If, by some impossible miracle, we had discovered genetics before anthropology and all those other disciplines, I wonder if scientific racism would never have emerged. Evolution deceives our eyes; it presents people as being similar when the underlying code says something different.

Brains are biological and therefore built upon genes, which vary between people and populations. Does the cultural, social, ancestral, and familial categorization of being a Jew have a biological basis that also renders Jewish people of apparently greater cognitive abilities than non-Jews, as is frequently asserted? Is the alleged gap between ethnically Black and other populations in

IQ testing rooted in genetic differences, or is it in our societies? Is the success of Jews at ostensibly intellectual activities such as chess, classical music, and science a result of a biological advantage over and above a purely cultural interest in those pursuits?

These two examples, physical prowess and intelligence, are a recapitulation of views that were aired at the birth of genetics as a discipline, a century ago, when racism was far more culturally acceptable. Arguments to support people's casual observations sometimes take this form: "Jews are good at intellectual pursuits because their own history of persecution and association with financial businesses over centuries has rewarded and bred superior cognitive abilities into them." Similarly: "Centuries of enslavement have bred physical power into Black people, which accounts for their success in certain sports." These are both potentially testable scientific hypotheses in the current era of genetics, though neither idea is new. People have been writing about Jewish brains and Black brawn for centuries, ever since the advent of anthropology, evolution, and a more formal study of biological inheritance in the nineteenth century, under the guise of science. These beliefs are common, and not at all exclusive to White supremacists. Here, we will test them.

My subject has a dark past, rooted in colonialism, in White supremacy, and in persecution. My own academic ancestry is intrinsically linked to the birth of scientific racism, to eugenics, and to the greatest atrocities in human history. These are stories and theories that desperately need revisiting, and we must revisit them armed with a twenty-first-century understanding of biology.

Genetics is woven into the history of race in every conceivable way. I will be forensically unpicking what genetics says about skin color today and in our history and ancestry, about intelligence, about our bodies and sporting prowess, about the myths of race, racial purity, and racial superiority. Moreover, this book is a tool—a weapon—to be brandished when science is warped, misrepresented, or abused to make a point, or to justify hatred.

PART ONE

Skin in the Game

Of all the racial signifiers humans use, skin is the most striking—so let us begin with color. Humans are a highly visual species, and pigmentation is the first and primary indicator that we fall upon to categorize people. Skin color is determined by genes, aside from the marginal effects of the sun.

Genes encode proteins; proteins enact biology, meaning that all life is made of or by proteins. Hair is made of keratin, which is a protein. The melanin that pigments hair and skin is not a protein itself, but its production is heavily under the control of proteins, which are encoded by genes. Though we all share the same set of genes, they are the same but different. Minor variations between two people in the sequence of a gene will manifest as subtly different proteins, and that makes the biological difference between all humans—different ways of spelling the twenty thousand or so genes that each of us has: color, or, for a Brit like me, colour.

We are confident in our understanding of the foundations of genetics, but linking the basic genetic code to the shape and function of a protein is tricky. As we are discovering more and more in the era of genomics, it is never easy, and mostly impossible, to predict the physical manifestation of the gene that encodes it—the phenotype (the outwardly observable traits) from the genotype (the underlying genetic makeup that determines those traits). In the nineteenth century, the scientist Gregor Mendel crossed pea plants together by the thousands and worked out that traits are passed from generation to generation in discrete patterns with strict rules. After the rediscovery of Mendel's work at the beginning of the twentieth century, the concept of a gene was defined as the unit of inheritance—a discrete bit of heritable information. In fact, this idea is much older, though it was codified scientifically only in the twentieth century. The earliest description of a genetic disorder comes from the Talmud in the rabbinic instruction that excuses some boys from circumcision in their first few days, if other male family members bled to death during the same procedure—exhibiting what we now know was hemophilia. That pattern of inheritance, as in the shape or color of Mendel's pea plants two thousand years later, is predicated on rules that are undeniably correct, and we call them Mendelian.

The picture of genetic inheritance turned out to be much more complicated in humans than in peas. Our old simplistic models of how a specific gene relates to a particular characteristic have been eroded in the last couple of decades. This is not news in relation to complex human traits, such as intelligence or diseases such as schizophrenia, where dozens or sometimes hundreds of genes have been revealed to play a small but cumulative role in their development. We've known this for some years. Genomes are complex and dynamic ecosystems, in which genes have multiple jobs in the body, depending on where and when they are required. A gene involved in the growth of an embryo just after conception might have a very different role later in life, or no role whatsoever. A gene may have multiple roles—an effect we call pleiotropy. Another phenomenon, known as epistasis, means that the impact of one gene is dependent on others; its effect can be positive or negative and can occur between completely different genes in networks, or even between the two copies of each gene that we all have, one set inherited from each parent. Genes do many things in many ways, and even over a lifetime of studying them, you will still find new ways the human genome works. The genetic code has remained static for billions of years, but evolution has incessantly tinkered with how it is used to build a life.

The textbook examples we use to cover the basic principles of biological inheritance are often ones that are concerned with pigmentation, such as with eye color, but they turn out to be not nearly as simple as we teach. We learn at school that blue and brown eyes are encoded by different versions of the same gene (referred to as an allele; the brown allele is dominant over blue, meaning that in order to have blue eyes you need to inherit a blue allele from your mother and father, and the presence of one or two brown alleles will give you brown eyes). That's true-ish, but is complicated by the fact that there is a gene involved in green iris pigmentation, and at least a dozen other genes have been shown to have an effect on eye color. The result of this network is that contrary to what we learn in school, it is possible for a child to have *any* color eyes despite the color combination of the parents' eyes.

Another example that has been a stalwart in the maintenance of a straightforward Mendelian model for inheritance is *MC1R*, a gene involved again in all pigmentation, but most obviously in the very visible trait of hair color. There are many variants of *MC1R*, but around seventeen of them change the behavior of the protein it encodes, causing it to produce a specific and unusual type of the pigment melanin. If you have two copies of one of these variants, you have red hair. In that sense, red-hairedness is a classic recessive trait: Only people with two red alleles of *MC1R* will have ginger locks.

That was textbook until December 2018, when a large genetic survey revealed that the ginger variants in *MC1R* account for around 70 percent of ginger-haired people, and that the majority of people with two supposedly ginger variants in fact have brown or blond hair. Almost two hundred genes appear to have some influence over pigmentation in hair, which is about 1 percent of the total number of genes in the human genome. It is only in the era of huge genomic datasets that this type of result could be exposed: The scientists responsible for the study looked at 350,000 people to reveal that the once-simple model of ginger hair is much closer to being inscrutably complex.

Throughout the short history of genetics, we have clung to simple models that explain seemingly simple traits, such as eye and hair color. But look at people's eyes, and you will see a full spectrum from the palest blue to almost black and, on top of that, mixed patterns within the iris, flecks of different shades and full-on heterochromia, where eyes can have clear sectors of different colors or, in some cases, each eye is a different color.*

---

* Full bilateral heterochromia is striking and beautifully rare, and is mostly caused by genetic mosaicism, that is, the condition of having two different sets of genes in different cells. Though it is often said that David Bowie had heterochromia, he actually had a completely different condition called anisocoria, where one pupil is permanently dilated and is unresponsive to changing light, giving the appearance of two different colors, even though both were blue. He got this after he was punched in the face during a scrap at school over a girl.

Attempts to categorize humans by such seemingly simple traits are not easy, and the underlying genetics wickedly complicated.

Skin pigmentation is no different. Melanin is the primary pigment in skin, and its function is protection. More than a million years ago our ancestors in Africa began to lose their fur as they moved to a life on open savannas rather than in woodland or jungles. Dense hair is hot, and they evolved new strategies for staying cool, including better perspiration and the loss of most of their body hair: We traded hair follicles for sweat glands. But this new exposure increases the risk of folate deficiency; that is, destruction of one of the key vitamins by the sun's ultraviolet rays. The result of this depletion is a whole suite of serious medical problems, including anemia and spinal defects during development in the womb. These are significant evolutionary pressures, and skin adapted to cope.

Specialized cells at the base of the skin called melanocytes produce melanin, which gets deposited in tiny packets—melanosomes—that migrate toward the light and sit atop other skin cells. In doing so, they simply absorb and block the UV rays before they can deplete folate levels in cells beneath. If you are pale skinned, you have less melanin, and therefore a reduced capacity to absorb UV in this way, so if I could offer you only one tip for the future, sunscreen would be it.

While these basic principles are understood, the picture is complicated by the fact that there are several types of melanin, whose production varies according to the cell's location in the body. Pheomelanin is a version that is pinker and features in red hair, the nipples, penis, and vagina. Eumelanin is more common and is found in skin, the iris, and most hair colors. Many genes are involved in the biochemical pathways that result in melanin production, and natural variation between people in the genes is the root cause of the spectrum of skin tones that humans have. Melanosomes vary in size and number between and within people, and this also influences visible pigmentation. Just like eye color, hair color, and most human characteristics, the genetics of pigmentation is complex, confusing, highly variable, and only partially understood.

It is wholly unsurprising that, on a continent of over 1.2 billion in fifty-four countries, the skin color of the peoples of the African continent is a vast tapestry, which overlaps with Indians and Aboriginal Australians, South Americans, and some Europeans. Yet we talk about "Black people" or "Brown people." The pigmentation of a pale-skinned redheaded Scot is a long way on a color chart from that of a typical Spaniard, though we call both of them "White." The skin color of more than a billion East Asians is similarly variable, yet nowadays, we tend not to refer to them by skin color at all. "Yellow," though an integral part of the description of East Asians for several centuries during the development of scientific racism, has fallen out of usage and is now generally accepted as being entirely inaccurate and simply racist. Instead, the main racial signifiers for East Asians are the epicanthic fold of the upper eyelid (which is also present in Berbers, the Inuit, Finns, Scandinavians, Poles, American Indians, and people with Down syndrome) and thick, straight black hair. Traditional racial categories are not consistent in their taxonomic boundaries.

Over the centuries, science developed and the process by which we apply taxonomic principles to humankind became more and more refined. Ultimately, human origins and human diversity would be scientifically unified in genetics. But they have always been considered together, at least since the seventeenth century, when we saw the first of many attempts to formalize what race is and how many races there are. In discussing the history of human classification, it is important to recognize the culture in which these descriptions were being undertaken, and although many are unpalatable and unscientific today, we can describe them as racist rather than necessarily condemning them as such.

There are plenty of references to skin color from ancient history, notably from Egypt, whose geography around the leviathan Nile runs a long way north to south, and thus encounters a range of skin tones according to their proximity to the equator—from the Mediterranean to the north to what is now the Sudan to the south. There is little evidence for class or social structure relating to skin tones in ancient Egypt, though variation is acknowledged in their art.

The Greek city-states and later empire stretched far and wide and was predominantly bound to the sea. They had many terms relating to identity, ethnicity, and nationhood: *ethnos*, *ethos*, *genos*, and others. There are also plenty of references in Greek literature to skin tones and pigmentation, though direct translation is not always straightforward. Their international reach extended from the east, and well into Africa. The earliest references to Ethiopia are in *The Iliad* and *The Odyssey*—the word itself is a compression of *aitho* and *ops*: "burnt" and "face." In *The Iliad*, Achilles's hair is described as *xanthos*, which might mean blond, brown, or even ruddy. As with all languages, ancient words don't necessarily map directly onto words in the present. Sometimes these words are used as descriptions of temperament as well as physical appearance, as in modern English—"blonde" becoming a derogatory term toward women to mean ditzy; "swarthy" is listed in some dictionaries as meaning saturnine or mysterious, as well as dusky. Odysseus is sometimes xanthos but also Black skinned at times, and in the translation of *The Odyssey* by Emily Wilson he is tanned. He was, after all, a complicated man.

Maybe to the majority of us who are not classical scholars, the assumption of Whiteness of the ancient Greeks stems from ancient statues that we see today as pure marble white, but which were brightly painted in their time. In contrast, depictions of men on ancient pottery were, for centuries, monotone black, though no one assumes that this meant that all Greek men were dark skinned.

Similarly, Rome was a huge intercontinental domain as both republic and empire. It enslaved people from the north and south, but also integrated non-Romans into society outside of servitude. That there were people from across Africa and the Middle East in Roman Britain is entirely uncontroversial. Knowing the proportions of these international groups is not easy, not least because of the diversity within the Roman Empire, and the lack of import or clarity of words used to describe pigmentation. Nevertheless, written and archaeological evidence is unequivocal. The second-century Roman governor of Britain, Quintus Urbicus, was born in Numidia, what is now Algeria. A gravestone from around the same time in South Shields marks the death of a woman from just outside London called Regina, a freed slave who married her former owner, a man named Barates, from Palmyra in Syria. In the genomic age, we can use DNA to assess the mixing between diverse groups from history, but so far there is a paucity in relation to Roman Britons. There are plenty of reasons for this: The net for surveying the genomes of Roman bones has not been cast wide yet, and it's quite possible that these genes have drifted out of contemporary genomes. Maybe there wasn't a great deal of sexual relations with the locals, what we broadly call admixture—there is similarly little trace of Danish DNA in Britons today, despite several centuries of comprehensive rule and Danelaw in the Middle Ages. Nevertheless, there are some clear indicators of African admixture. In 2007, a small cluster of White men from the north of England with no known connections to Africa were shown to bear Y chromosomes that are most commonly found in countries such as Guinea-Bissau, and this gene flow may have occurred in Roman Britain.

This is not to paint a picture of Europe's past being a utopian melting pot of equality. Far from it—these were times of extensive slavery and colonial expansion. Religious and ethnic stereotypes and prejudices abounded. But their criteria for subjugation were not the same as ours today, and pigmentation has not always been a primary determinant of character or descent.

Within Islam in the Middle Ages, there is minimal discussion in the surviving literature of superiority or prejudice based on skin color, until the writings of the eleventh-century philosopher Avicenna, who believed that people exposed to extremes of climate (relative to the Middle East) were more suited to slavery owing to regionally determined differences in temperament: Pale-skinned Europeans were ignorant and lacked discernment, dark-skinned Africans were fickle and foolish. Both, therefore, were suited to oppression during a period that encompassed more than nine hundred years and upward of five million enslaved people.

The emergence of a scientific (or, more accurately, pseudoscientific) approach to human taxonomy coincided with the growth of European empires. Characterization of different populations before the expansion of Europeans around the globe was more likely to be based on religion or language than skin color, but with the birth and growth of the era of scientific revolution, pigmentation became essential to the character of humans.

While it is true that some of the pioneers of anthropology had scientific principles at heart, the othering of people in potential or actual colonies has the effect of permitting subjugation. It is far easier to sell the case for occupation and enslavement if you are persuaded that the indigenous people are different, have different origins, and are qualitatively inferior to colonists. Despite this process of radicalization, from the seventeenth century, some adhered to a Christian view that was less racially divisive, as all humans were children of Adam and Eve. This idea, monogenism, was supported by such significant scholars as Robert Boyle and Immanuel Kant. Kant formulated an idea that there was a single origin for humankind, but that fixed differences, primarily in skin tone, emerged from local conditions.

The opposing theory, polygenism, claimed that distinct human populations arose in the areas they currently inhabited, and hence had different biological and cultural behaviors via isolated evolution. Supporters included Voltaire. That voice of Enlightenment thinking was an ardent polygenicist, writing in 1769 that:

> Our wise men have said that man was created in the image of God. Now here is a lovely image of the Divine Maker: a flat and black nose with little or hardly any intelligence. A time will doubtless come when these animals will know how to cultivate the land well, beautify their houses and gardens, and know the paths of the stars: one needs time for everything.

The Swedish naturalist Karl Linnaeus founded the taxonomic classification of all living things that we still use today: genus and species—*Homo sapiens.* In 1758, in the tenth edition of his classic *Systema Naturae*, he included us in five categories or subspecies: *afer* (meaning African), *americanus*, *asiaticus*, *europaeus*, and *monstrosus.* A big part of his scheme was skin color, but it is worth noting that he applied all sorts of racist value judgments on top of the more prosaic biological traits: *Afer* were lazy, cunning females without shame and ruled by caprice; *americanus* were red skinned, with straight black hair, and were zealous and stubborn and ruled by customs; *asiaticus* were severe, haughty, greedy, and governed by opinions. And his judgment on the subspecies *europaeus*? Gentle, acute, inventive, and governed by laws. As for *Homo sapiens monstrosus*, Linnaeus mixed legend with contemporary science, including mythical and somewhat bizarre humans: feral people, wolf boys and wild girls, Patagonian dwarfs, and single-balled Hottentots.

Not everyone was quite so antagonistic and racist in their attempts at categorizing people and justifying racial hierarchies. In the eighteenth century, the German anthropologist Johann Blumenbach was one of the first to apply scientific principles to populations. He also put humans in five taxonomic ancestral groups: Caucasian (meaning White Europeans); West Asian and North African; Ethiopian (meaning sub-Saharan Africans); Mongolian (that is, East Asians excluding Southeast Asia, which he categorized as Malayan); and American Indian. Craniometry, based on measurements in sixty skulls, was the big part of his scheme, much more than skin color, though some of the pigmentation epithets that are still used today come from this taxonomy: He referred to his five categories as White, Black, Yellow, Brown, and Red. As a particular sort of biblical creationist, he argued that Adam and Eve were White-skinned Caucasians born in Asia, and their descendants had migrated from this base all around the world. This was known as the degenerative hypothesis, that races emerged from local environmental conditions, as in the case of darkened pigmentation in Africans as a response to the sun.* Within this scheme, Blumenbach was adamant that these five varieties were all one species.

It's interesting to see that Blumenbach was close to modern scientific understanding of human migration and evolution, though wrong in almost every facet. We now know *Homo sapiens* to be an African species in origin, probably pan-African, with roots certainly from the Rift Valley in the east, but also in northern Africa, where the oldest remains of our species have been found, dating to around three hundred thousand years ago. We know that pale skin is an adaptation via the process of natural selection to exposure to a weaker sun in cloudier northern climates. It's also worth noting that Blumenbach was more restrained about asserting that Africans were inferior to White Europeans: "There is no so-called savage nation known under the sun which has so much distinguished itself by such examples of perfectibility and original capacity for scientific culture, and thereby attached itself so closely to the most civilized nations of the earth, as the Negro."

Another contemporary of Blumenbach—and antagonist of Kant—was Johann Gottfried von Herder, who took a scientific view that appears even more modern: He argued that the idea of four or five racial categories was specious. "The colors run into one another," he

*The meaning of "degenerative" is not necessarily the same as we would use today, but implies a change from an earlier form rather than explicitly a fall from a perfect inviolate form.

wrote, and saw human variation on a continuum that "belongs less to the systematic history of nature than to the physical-geographic history of humanity."

Herder's assessment is impressively in line with twenty-first-century scientific views on the global human journey.* His voice was lost to Kant's louder assertion that skin color was inherently linked to character, bound by biology, innate, and that it was therefore a legitimate way to categorize and rank humans. Those with pale skin were superior to those with dark.

Kant also was steadfast in his view that these qualities were immutable. The blackness of African skin was set in stone, and with it came stupidity, along with certain related characteristics. He shared that idea of fixity with the French naturalist Georges Cuvier, who opted for three races of humans in 1798: Caucasian, Mongolian, and Ethiopian. He ranked them in that order, too, the Caucasian race being the most beautiful and "superior to others by its genius, courage and activity."

---

*Though we should not make the mistake that Herder was either non-racist or the very model of a modern scientist. "A ministry in which the Jew is supreme," he wrote, "a household in which the Jew has the key of the wardrobe and the management of the finances . . . are Pontine marshes that cannot be drained." He also claimed that "a negro child is born White: the skin around the nails, the nipples and the private parts first become coloured . . . The projection of the mouth would render the nose short and small, the forehead would incline backwards, and the face would have at a distance the resemblance of that of an ape."

By the nineteenth century, biology was inching toward a true revolution. The idea of evolution and a move away from special creation was part of the changing scientific culture, and in 1859, Charles Darwin's *Origin of Species* would unveil the truth of the history of life on Earth, and the process by which all living things, us included, had come to be. In the years following the unveiling of natural selection, the continuity of life on Earth became the prevailing idea, though classification and taxonomy are still necessary: Life is continuous, but there are real and nonnegotiable boundaries between creatures. Classification within our species was something that Thomas Huxley, Darwin's friend and greatest defender, had a stab at in 1870, and though he stuck to the Linnaean concept of the "four great groups of mankind," as Huxley put it (now excluding the *monstrosus*), he went into far greater resolution, delineating dozens of individual populations and trying scientifically to account for their differences. He applied population subdivisions whose names mercifully never caught on, such as *xanthochroi* for fair Whites, and *melanochroi*, the darker-skinned Europeans toward the Mediterranean. Though he employed the technical jargon of a Victorian scientist, and relied heavily on skull measurements, Huxley also used indecipherably imprecise language: "The stature of the Negro is, on the average, fair, and the body and limbs are well made." But he also acknowledged the mixing of all these populations, a fact demonstrated by genetics in the twenty-first century.

And so it went on. In the twentieth century, the influential US anthropologist Carleton Coon outlined five classes of *Homo sapiens*: Caucasoid, Mongoloid (which included everyone indigenous to the Americas as well as East Asia), Australoid (meaning Aboriginal Australians), and two types of Negroid—Capoid and Congoid (from southern Africa, near the cape, and from the Congo). Contemporary science has rejected these classifications, though they definitely persist in some older members of the public, who have referred to these categories in question-and-answer sessions when I give lectures.

The continual failure to settle on the number of races is indicative of the folly of the endeavor. No one has ever agreed how many races there are, nor what their essential features might be, aside from the usual sweeping generalizations about skin color, hair texture, and some facial features. It is difficult to untangle the rationale, the evidence, and the motivation for the plurality of pre-Darwinian views of human origins.

"Time makes ancient good uncouth," the poet James Russell Lowell wrote. The archaic language in the writings of the eighteenth- and nineteenth-century anthropologists is not always clear in modern scientific terms—race and species are sometimes used interchangeably; some seem more scientifically minded, such as Blumenbach's, others such as the opinions of Kant and Voltaire are unequivocally and perniciously racist by today's standards. Every one of these ideas though must be considered in the cultural context and time in which it was authored. All are by European men being exposed to the peoples of the world as a result of expanded trade routes, colonialization, and empire building, and in many cases the conquering and enslavement of the people they encountered. The invention of race occurs in an era of exploration, exploitation, and plunder, an era when the othering of people from colonies extended to actual human zoos.

In 1810, a Khoikhoi woman called Saartjie Baartman was brought to London from Cape Town, where she was exhibited on stage in Piccadilly, sometimes on a leash: "the Hottentot Venus—the Greatest Phenomenon of the Interior of Africa." Saartjie (or little Sara) was her Dutch given name—her birth name is now forgotten. Sara's ethnicity was certainly part of the appeal, but she was presented in the same manner as other contemporary "living curiosities," extremes of obesity and skinniness, of height and of other medical abnormalities, what we would today call a freak show.

After four years in London and on tour around the UK, Baartman was sold to a French animal trainer and exhibited in the Palais-Royal in Paris. There she lived effectively as if enslaved, and scientists inspected her, including Georges Cuvier. Particular interest was in a

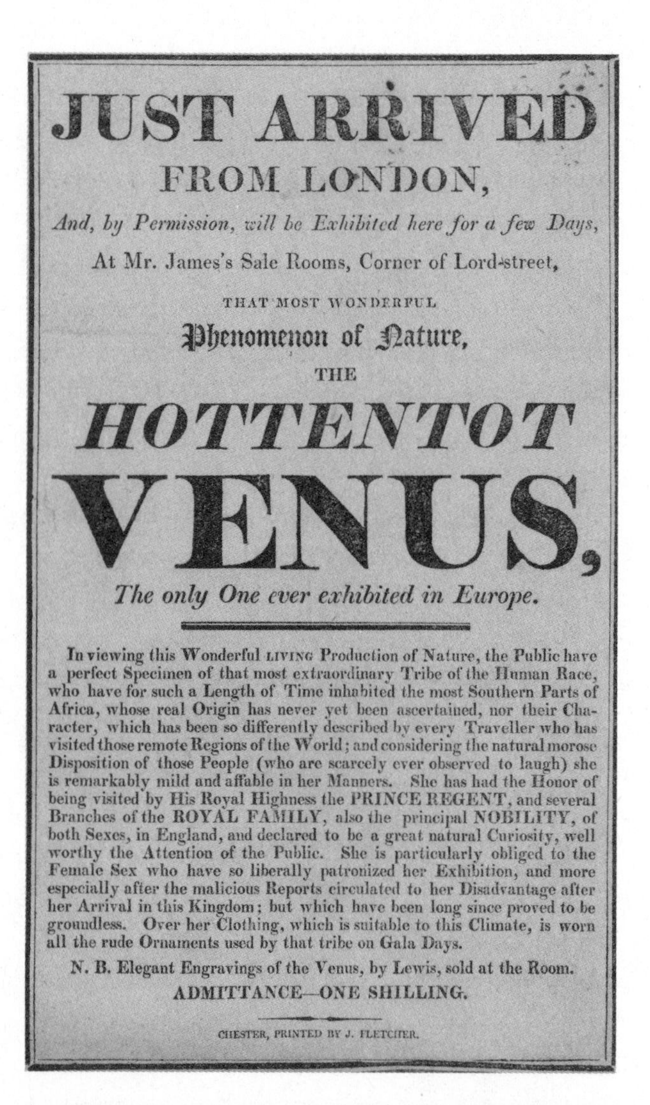

An 1810 poster touting the arrival of the Hottentot Venus (presumably Sara Baartman), "a perfect Specimen of that most extraordinary Tribe" from "the most Southern Parts of Africa"

common feature of the Khoisan people called steatopygia, fatty deposits on Baartman's buttocks and breasts, and in the anatomy of her labia, which, though never publicly revealed, were thought to be comparatively enlarged. Sara Baartman died aged twenty-six in 1815, possibly from smallpox or syphilis. Cuvier conducted her autopsy, though not to assess the cause of death but as a further inspection of fundamental anatomical features. This grim tale of exploitation and literal objectification is a key part of Cuvier's developing ideas of scientific racism, her body assumed to be typical and fixed for his category of Ethiopian—a class of human disconnected in behavior and history from other, inherently superior breeds.

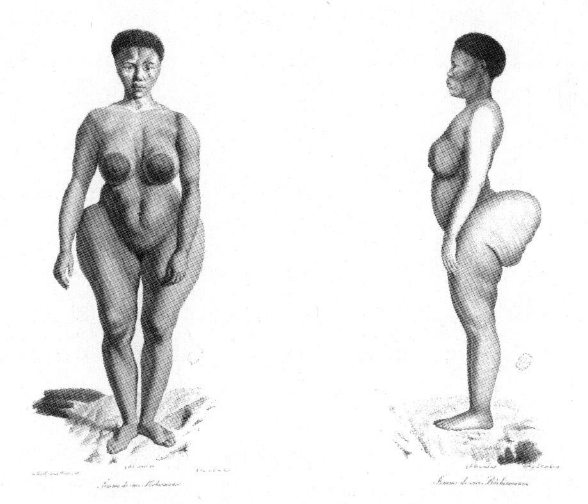

Images of Sara Baartman from the pages of *Illustrations of the Natural History of Mammals*, which were sandwiched between entries on a type of wild sheep and a langur monkey

There was a gradual move away from polygenism in the nineteenth century. Darwin's theory of evolution was built upon the antiquity of humankind and the age of Earth being millions of years rather than the standard six thousand believed by creationists. Nevertheless, well into the second half of the twentieth century, scientific debates about human origins set the Out of Africa hypothesis against the Multiregional hypothesis. Did *Homo sapiens* evolve in Africa and then disperse around the world, or had an earlier ancestor of ours left Africa much earlier, and did the differences we see in contemporary populations evolve as distinct lineages?

This was an interesting twentieth-century recapitulation of monogenism versus polygenism, though it should be stated that the Multiregional hypothesis was not ideologically racist—it was just wrong. By the 1990s, with a veritable catacomb of ancient human fossils found in and around the Rift Valley, Out of Africa had definitively won. A model of the spread of *Homo sapiens* around the world now relied upon its origin in Africa. This is now universally accepted, though there are plenty of details that will still emerge.* As mentioned, the oldest members

---

*In July 2019, evidence of a *Homo sapiens* skull was described from a cave in Greece and dated to 210,000 years ago. This was not the first evidence of earlier dispersals from Africa by our species, but it's the oldest, revealing a dispersal much earlier than previously known. We do not see continuity into the present of these earlier migrants and so presume that their lineages went extinct, possibly replaced by Neanderthals.

of our species (albeit more archaic forms) resided in what is now Morocco, not in the east of Africa. And all the evidence firmly points to there being an Out of Africa dispersal some seventy thousand years before today.

This deep prehistory of our species is intrinsically important from the point of view of the history of racial taxonomy. Anthropological sciences had begun to merge with new biochemical techniques at the beginning of the twentieth century, a trajectory that would be fully realized with genetics in the twenty-first. The biology of difference was about to get molecular, and it began not with skin but with blood.

The idea of blood being the carrier of inheritance is old. We talk of bloodlines, and pure blood, but DNA and genes are now equivalent in terms of colloquial descriptions of inheritance. Blood plays a significant role in twentieth-century studies of human categorization, too. The ABO system was described at the beginning of the twentieth century. The different blood types reflect subtly different alleles of the *ABO* gene, and this was the first time this type of genetic difference had been described. In 1919, Ludwik and Hanka Hirszfeld looked at these types in sixteen different groups of soldiers to see if they varied by nation (they included Jewishness among these). The husband-and-wife team found A and B types were distributed all over the world in clusters. This provided the basis for their theory that there were two historical races of humans who subsequently mixed, which explained why similar blood groups were found a long way apart. In fact, the same ABO blood system is found in gibbons and Old World monkeys, and predates hominid lineages. The Hirszfelds couldn't quite hide their own prejudices in devising these tests though: "It was enough to tell the English that the objectives were scientific," Ludwik Hirszfeld wrote in his autobiography.

> We permitted ourselves to kid our French friends by telling them that we would find out with whom they could sin with impunity. [When] we told the Negroes that the blood tests would show who deserved a leave, immediately, they willingly stretched out their black hands to us.

The oft-quoted classic study of genetic diversity by Richard Lewontin also used blood to test concepts of race. In the 1972 paper "The Apportionment of Human Diversity," Lewontin found that the vast majority (85 percent) of genetic differences were *within* classical races, not between them. Only 6 percent of differences segregated by race. This conclusion has been questioned on and off since its publication, but remains broadly correct. The main challenge was formalized as "Lewontin's fallacy" in 2003 by the mathematician Anthony Edwards, which pointed out that if you aggregate multiple sites of variation across a genome, you can in fact predict the population from which a person comes accurately. Both results are true; it just depends on the detail and the resolution.

And so, as we got better and better at reading genomes, and applied our knowledge to more and more of them, we were able to make pictures of population difference at increasingly high resolution. One of the great studies in the twenty-first century of human population genetics came in 2002, in the early years of the genomic revolution. By taking the genomes of many people from around the world, we were now in a position to dip into all those genomes and ask how similar they are. This technique is reliant on being able to sample individual letters of difference between people assumed to be representative of a population, and then asking a computer program to cluster them together, almost as if creating a map of similarities. Noah Rosenberg and his team analyzed 1,056 people from 52 geographic regions, and looked at 377 places in the genome where the DNA there is known to vary between people.* With this particular technique, you ask the computer to gather these variables into a set number of clusters. When you start with two, it identifies one group of humans as African and Eurasian, and the other as East Asian, American Indian, and indigenous Australian. With three clusters, Africa is split off as a separate group. At five, indigenous Australian becomes a separate group alongside African, European (including West Asian), East Asian, and American Indian.

---

*The regions in this study are called microsatellites, and are short stretches of DNA that repeat like a stuck record. It's the *number* of repeats that varies between people.

That is remarkably similar to the classic racial taxonomies from the era of scientific racism. Does this mean they were right after all? Well, no. This type of analysis shows broad similarities in populations: It reflects geographical landmasses, which are not insurmountable barriers to reproduction but do hinder interbreeding; it also reflects evolutionary history and migration. The data also showed long, clear gradients between all the clusters, and no unambiguous way to say where one cluster ends and another begins. Without sharp boundaries between these population structures, instead it showed continuity between people. This, as Johann Gottfried von Herder had suggested, was because human variation will not succumb to an imposed artificial taxonomy but instead reflects history.

Rosenberg's paper is often used by racists to erroneously claim that there are indeed five genetically distinct races. In fact, it does no such thing, and this is obvious in the data: When the clusters are set at two, Africa, Europe, and West Asia are lumped together as one and the rest of the world as another. There is no a priori reason to settle on five clusters as being the definitive categorization of humans, and deciding to do so because it corresponds with an earlier yet debunked classification is simply affirming preexisting biases. When you increase the cluster number to six, the next distinct group to emerge is the Kalasha. They are a northern Pakistani tribe of around four thousand people who marry almost exclusively within their own ethnic population, which is tucked away in relative isolation in the mountains of the Hindu Kush. Though these people are somewhat genetically distinct, not even the most committed racist describes the Kalasha as a sixth human race.

Bear in mind that all these studies rely on complex statistical analyses on ever increasing datasets, and that they are based on genotype, not phenotype. What that means is that even if the differences and similarities in DNA make useful proxies for predicting the populations from which they were sampled, they don't necessarily correlate with the traditional categories of race, as determined primarily by pigmentation. This type of analysis is totally valid, and is the basis for studying human history, migration, and genetic variation between populations and people. We could keep increasing the number of clusters and find ever more precise similarities and overlaps. When even higher-resolution genetic mapping was applied to the people of Britain in 2015, families who had lived in the county of Devon for multiple generations could be distinguished from the people of the directly adjacent county of Cornwall, and when these precise differences were plotted on a map, the boundary was the River Tamar, which for centuries has effectively been the county line.

When the same technique was applied to the Iberian Peninsula in 2019, it showed vertical stripes of similarity, revealing that, because of the peculiarities of Spain's history, people are fractionally but measurably more similar in a north–south axis than they are east to west.

Similar studies of the underlying population structure of the US also show something of its recent history of migration. In the Midwest, we can see traces in genomes that bear similarities with Finns and Swedes. We can genetically detect clusters of people in New Orleans who are similar to Acadians—French-speaking Atlantic Canadians. In the eighteenth century, Acadians were expelled from their homes and eventually settled in Louisiana, and their name "Acadian" mutated into "Cajun."

Are the peoples identified in any of these studies functionally different? Of course not, it's simply that we have become so good at identifying the genetic histories of populations that we can pick up these wispy, diaphanous traces of similarity and difference. We could eventually cluster all humans into seven billion individuals, because every human genome is unique.

Humans suffer universally from a syndrome that Richard Dawkins called the "tyranny of the discontinuous mind." We yearn to categorize things and fail to recognize continuity. We strain to put things into discrete boxes, and define things by what they *are* rather than what they *do*. This is a problem in science, and one that relates to the Linnaean classification that biologists cling to. Linnaeus was looking to a system that reflected inviolate platonic forms of creatures (and rocks—it was his classification that gave us the silos of animal, vegetable, and mineral, which, while being the basis of twenty questions on long car journeys, is not a very good way of classifying living things). Contemporary thinking during the era of colonial expansion, and later the Enlightenment, was primarily that biblical creation was the story of humans, and classification of the people of the world was derived from models that reflected a single origin, or degeneration from that form in God's image. Even though monogenists recognized an early form of regional adaptation, and evolution was in the air as a concept, it was only with Darwin's mechanism of natural selection in 1859, and then applied to humans in 1871 in *The Descent of Man*, that the natural histories of humans could be explained. In the genomic era, our data will continue to show not discrete classification but the immensely complex story of human life on earth over hundreds of thousands of years of prehistory, and a few thousand years of history.

That historical monogenist view is now known to be correct in principle but wrong in every detail. *Homo sapiens* is a creature whose origins are in Africa. There were dispersals out of Africa and into Eurasia in the last 210,000 years that petered out, and those people leave no genetic legacy in living humans, as far as we can tell.* The main emigration from the true motherland occurred around seventy thousand years ago, and these people, maybe numbering only a few thousand, would form the population from which the rest of the world would primarily be drawn. This is evident in the bones of our ancestors, and in our living genomes. But let's be clear on the messiness and timescales we are talking about here. Out of Africa was not an "event" as we think of it, nor was it a migration in modern terms either. The rate of passage is over thousands, if not tens of thousands, of years, and indeed we have genetic evidence of back migration within the last few thousand years, too. So while a population was established in a previously uninhabited location, that is not to say the gate closed behind them. They were not setting out to conquer a promised land, just meandering over generations, on average, away from the continent of Africa.

So here is the baseline: All humans share almost all their DNA, a fact that betrays our recent origins from

*The discovery of earlier migrations out of Africa is sometimes used by racists as evidence that Europeans are such a distant branch from Africans that they are effectively two different species. This, it almost pains me to say, is a really dim-witted argument. We see no evidence of any living descendants genetically from earlier Out of Africa dispersals, nor do we even see permanence in *Homo sapiens* in Europe until their arrival within the last fifty thousand years. This argument is so devoid of reason or fact that it's not even wrong, and it makes me wish racists actually had better arguments.

Africa. The genetic differences between us, small though they are, account for much, but not all, of the physical variation we see or can assess. The diaspora from Africa around seventy thousand years ago, and continual migration and mixing since, means that we can see that there is structure within the genomes that underlies our basic biology. Very broadly, that structure corresponds with landmasses, but within those groups there is huge variation, and at the edges and within these groups, there is continuity of variation. Of all the attempts over the centuries to place humans in distinct races, none succeeds. Genetics refuses to comply with these artificial and superficial categories. Skin color, while being the most obvious difference between people, is a very bad proxy for the total amount of similarity or difference between individuals and between populations. Racial differences are skin-deep.

We are now in the era of ancient DNA, where the fragmented genomes of creatures long dead can be wangled out of teeth, bones, and even the earth where they died. The first great headline in this new old world was the resurrection of *Homo neanderthalensis*—the Neanderthals—when, in 2009, a partial genome of a man who died in a cave fifty thousand years ago was reassembled. Since then, dozens of other dead or extinct human genomes have been jigsawed back together, and the story of human evolution has been radically transformed as a result. New types of humans have been identified from DNA taken from bones that themselves were not enough to allow classification. We can now piece together tales from our shared past that were otherwise lost in time.

Often, geneticists asking these questions of the ascent of humankind focus on how the sequence of DNA itself changes over time and space, and pay less attention to what phenotype might have emerged from the genotype—as if we were studying sheet music without considering what it sounded like. But it's interesting to think about what those ancient people were like. Here we face the ongoing problem in human genetics—that it is not at all easy to extrapolate phenotype from genotype. As mentioned earlier, you might have two copies of a gene that we had thought ensures red hair, but most people with that genotype do not have ginger locks. We are on more solid ground when it comes to things related to diet; for example, genes that relate to high-fat diets tend to occur at a greater frequency in people whose diet includes a lot of fish or seafood, such as the Inuit, and we can see that these traits have been selected as local adaptations. We can see the genes that allow milk drinking in White Europeans and a few clusters of dairy-farming pastoralists dotted around the world.

But we are very visual in our thinking, and we would all like to know what these people looked like. Old bones tell us a lot, and we can infer simple things like height and stature with careful reconstruction, and even much more subtle physical traits, such as whether people were left- or right-handed, based on bones thickened through use and the shadows on those bones of heavier musculature. Neanderthals were robust, big-chested, and muscly. Some researchers think that maybe this physical presence suited them to sprint running and ambush hunting. This fits with a life in wooded lands, springing traps or jabbing spears into mammoths, wild sheep, or boars. Indeed, the genetics may reinforce that picture—some research suggests that they had more versions of genes that today we associate with explosive energy rather than stamina (though as we will see in part 3, there is plenty of controversy about the importance of these genes in athleticism). There's also disagreement about the value and accuracy of facial reconstructions; whether such a reconstruction actually resembles the person in life is frequently contested, and as far as I know, the test of this hasn't been done: a reconstruction of a living person based on a scan of the person's skull.

When it comes to pigmentation, we are in even more treacherous waters. Eye color genes are plentiful, and if you want to know what color eyes a long-dead person had, we can give you probabilities, not answers: My report from the commercial ancestral company 23andMe tells me that 31 percent of people with the same version of the gene *OCA2* as me have dark brown eyes, which means that 69 percent do not, and 13 percent have blue or green eyes. I have dark brown eyes—I know this because I own a mirror; if aliens were to dig me up in fifty thousand years and extract my DNA, on current knowledge what are the chances they'll get my eye color right?

Skin is even tougher. Pigmentation is not a binary trait, even though we use binary terms like "Black" and "White." More and more we are discovering that genes play multiple roles and have many interactions with other genes in complex metabolic pathways. The traditional anthropological view has been that humans in Africa before the great diaspora were dark skinned as an adaptation to the hot sun. Lighter skin evolved probably in response to colder, cloudier latitudes, as described earlier. The traditional genetic view has been that there are a handful of genes—maybe fifteen—that account for the majority of the differences we see in pigmentation, suggesting a relatively straightforward genetic architecture. However, this is at odds with some observations. A peculiarity of the genes that influence pigmentation is that while we see natural selection at play in the broad sweep of skin color at changing latitudes, it does not account for the differences we see in pigmentation at the same latitude. It is simply not the case that everyone who lives on the equator has the same darkness of skin. Nor is it true that the Inuit, Iñupiat, Russians, Finns, Icelanders, and everyone else who lives at the sixty-sixth parallel north have identical skin tones. Obviously and significantly, other factors are at play, aside from pigmentation in relation to sunlight.

We can see the effect of particular alleles of genes such as *SLC24A5* and *OCA2* (and a few others) in lightening the skins in European and Asian populations, and these important adaptations have dominated our thoughts about the evolution of pigmentation. But as in so many domains of science, we have until very recently virtually ignored the continent of Africa. There is more genetic diversity in Africa than the rest of the world combined. What this means is that there are many more points of genetic difference between Africans than between Africans and anyone else in the world—two San people from different tribes in southern Africa will be more different from each other in their genes than a Briton, a Sri Lankan, and a Māori. And there is more diversity in pigmentation in Africa than in the rest of the world, too. Only in the last few years have researchers begun to study the genetics of African skin, which is somewhat ironic given that five centuries of racism have been almost entirely based on it.

The picture that is beginning to emerge is truly shaking things up. In 2017, the geneticist Sarah Tishkoff led a team that sampled DNA from more than 1,500 people from Botswana, Ethiopia, and Tanzania, and also assessed the levels of melanin in the skin in their forearms. In making this comparison, they could associate genetic differences with skin tones. The most common variant was one in the gene *SLC24A5*. This variant is strongly associated with light skin but was found at high frequencies in Ethiopian and Tanzanian people. Obviously, it does not have the lightening effect in these people, but appears to have returned within the last few thousand years from Eurasia to Africa, where it is now common. Further variants in known genes and otherwise unsurveyed areas of the genome emerged out of this study, some associating with lighter and some with darker skin. This reflects the enormous intricacy of pigmentation genetics, but more interesting is that these variants all appear to have been present in our genetic lineage for hundreds of thousands of years, that is, before the evolution of *Homo sapiens*. Tupac rapped that the darker the flesh, the deeper the roots. That alas is not correct. The idea that we were ancestrally dark skinned before diversifying as we crept around the globe is now known to be incorrect. Not only were we diverse in our skin color long before the dispersal from Africa, we were diverse in our skin color before we were our own species.

Yet another contemporary account of the complexity of pigmentation and the histories of humans was published in late 2018. The Khoisan peoples are noticeably lighter skinned than many other populations in southern Africa. They are quite distinct both genetically and in terms of skin tone, possibly reflecting a relative degree of cultural separation for thousands of years. But no population is ever completely isolated or static. Though the resolution of genomics in Africa is currently lower than for other parts of the world, we do know that there has been gene flow into Khoisan ancestors in the last few thousand years. This includes from pastoralists, maybe from Ethiopia or the Near East two thousand years ago, via the expansion of Bantu-speaking culture during the Middle Ages, and in the modern era via the Dutch from trading posts in the Cape. Geneticist Brenna Henn has worked closely with the Khoisan for many years, and has established that their lighter skin is associated with the gene *SLC24A5*. The most common version in the Khoisan is the same as in Europeans, and Henn's work shows that it was introduced from migration *into* Africa in the last two thousand years. That it has reached such a high frequency in the Khoisan so quickly is strong evidence of intense selection for lighter skin, which shows both the diversity of skin color within Africa and the recent admixture between people coming back into Africa in the last few millennia.

Pigmentation is complex. Not atypically complex compared with other human traits, but it is visible and important. I don't wish to give the impression that we understand the emerging picture of the intricacies of skin color, only that the previous picture is cripplingly simplistic. We are rightly interested in how, when, and why skin color has changed over time, and pioneering studies by the likes of Sarah Tishkoff, Nina Jablonski, and Brenna Henn are helping to explore cavernous gaps in our knowledge, not least by engaging with African people in ways that have not happened before.

We also now have access to the DNA of the long dead, and can try to piece together puzzles about racialized characteristics from the past. This is categorically more difficult than with living people for two reasons. The first, as mentioned before, is that establishing phenotype from genotypes is never simple: We cannot measure the skin tone of a woman or man and compare it to the DNA, as that is all that remains. The second reason is the paucity of samples. Genetics is a comparative science, emboldened by the genomes of more and more people. One genome is packed with information, but two is much more informative, and thousands is when you've got good game.

But them's the breaks. We work with what we have, and questions about the most obvious phenotypes in our ancient ancestors are important, even if they aren't representative of any kind of presumed racial categorization or history. In 2016, the DNA of an ancient Briton was sequenced and presented to the world in the form of a bust—a head-and-shoulders bust of a kindly faced man, with very dark skin, tightly curled black hair, and blue eyes. He is a striking image of a British person long before the Picts, Romans, Vikings, Angles, or Saxons bothered our coasts. The scientific paper that prompted the model was much more circumspect, describing the pigmentation of this man with due scientific restraint: "Cheddar Man is predicted to have had dark or dark to black skin, blue/green eyes and dark brown possibly black hair." The DNA evidence had shown that he lacked pigmentation alleles that are associated with light skin. In the reconstruction, he is very heavily pigmented, a skin tone similar to that of a Sudanese man, or equally someone from Sri Lanka. When these pictures hit the news, racists all around the world lost their collective marbles with splenetic fury. That there were dark-skinned people in Europe ten thousand years ago is not at all controversial, so objections to the mere presence of a dark-skinned Cheddar Man in Britain are of little concern: Diversity in pigmentation in Europe is a fact of prehistory. But the deep darkness of his skin was a choice by the artist, and some geneticists grumbled about it.

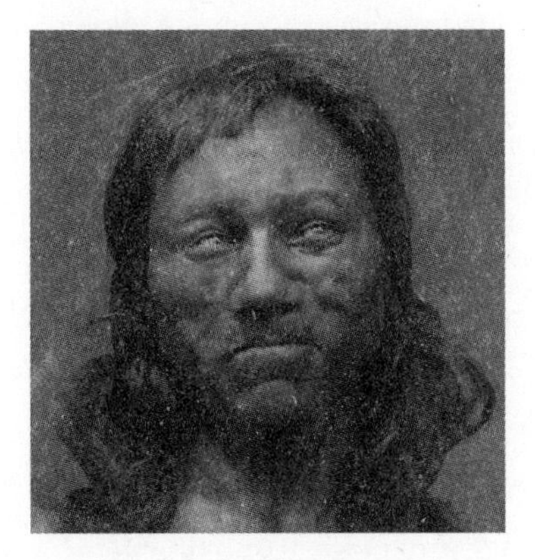

The bust of Cheddar Man

Based on multiple pigmentation genes, we think Neanderthals had moderately light skin, as did common ancestors of both *Homo neanderthalensis* and *Homo sapiens* half a million years ago or more. Some of the Neanderthal genomes indicate that they had a version of *MC1R* that was different from any seen in living people. The press started speculating that they were ginger haired, and in museums all over the world you will see vaguely redheaded Neanderthal mannequins. These *MC1R* variants have never been seen before, and biochemical attempts to get them to produce actual pigment in a petri dish were inconclusive. The pigmentation we see in reconstructions of Neanderthals on TV and in museums is speculative.

The truth is that it's very difficult to get right. Pigmentation changes during life, intrinsically—babies are not the same color as their adult versions—and extrinsically via exposure to the sun. Certain genetic variants predispose kids' pigmentation to change during their lives. I understand the need for people to see what our ancestors looked like; it is important to humanize prehistoric people, especially Neanderthals, who, far from the brutes of common lore, were sophisticated, cultured, artistic, and largely the same as us in terms of behavioral modernity. And with that comes the first cue, which is skin color.

We must be cautious. Brenna Henn told a meeting of geneticists and anthropologists in 2017 to "stop saying you can predict skin color from ancient DNA; you can't."

That message is pertinent for scientists trying to get a handle on the looks of our deep ancestors, but is even more significant when forensic DNA samples are used to predict the skin color of criminals. The most up-to-date science, collecting genes from the biggest samples of the most diverse people, makes it clear that DNA is a bewilderingly inscrutable predictor of skin color.

What we are beginning to see is that our deep evolutionary history was much less linear, with much more meandering and variance, than we had previously thought. As ever in biology, our attempts to instill the human story with a straightforward, comprehensible narrative have been thwarted by the inherent messiness, noise, and unruliness of our own evolution, combined with timescales and migration patterns that are almost unfathomable. We now know that pigmentation is a diverse spectrum and that it has been that way for hundreds of thousands of years.

What we can also say with an arsenal of scientific ammunition is that though skin color is the first and most obvious way we see humans, it's a superficial route to an understanding of human variation, and a very bad way to classify people. Our view of reality, so profoundly limited, has been co-opted into a deliberate political lie. We say "Black" when what we mean is "recently descended from a continent that has more genetic diversity and pigmentation diversity than anywhere else on Earth."

What we see with our eyes is the merest fraction of a human. The metaphor of a tree is how we tend to think of evolution, with its trunk and boughs, bifurcating into unique and discrete branches, until it gets to the twig of humankind. But the explicatory power of a tree for understanding huge swaths of human behavior and our evolutionary trajectory is seriously limited. Comparing human evolution to a tree works only if we mean trees that have been cultivated by us: pollarded—cut back to nurture new growth; espaliered—coaxed along other migratory pathways; and pleached—forced to entwine with other branches. That we are the product not of a tree but a tangled bank may not be obvious. But that is why we invented science: to free ourselves from the shackles of perception, to see things—including people—as they really are.

PART TWO

Your Ancestors Are My Ancestors

Family and ancestors are the ties that bind us to our pasts. Our immediate kin provide context for our lived lives: getting born, getting married, getting dead, or any version of that trajectory. You share half your DNA with each parent and a different half with each sibling (unless you are an identical twin, in which case it is close to 100 percent). These numbers go a long way toward explaining why you look more like your family than random strangers, and indeed behave comparably (a shared environment accounts for the rest).

Your family tree sits as an infinitesimally small node on the global tree of all life, tortuous and un-treelike as it is. Evolution is also a log of parents and children, and how they differ over oceans of time. In between these two levels of scale lies genealogy.

Ancestral belonging and genealogy are things that fascinate us all, but racists especially: Genealogy is possibly the second most popular pastime in the UK (after gardening), and the first in the US. Many of the arguments put forward by racists center around belonging to specific demographics, the othering of different groups, and the displacement of people. Many non-racists are also concerned with immigration in the modern era, but few express the sense of a people being replaced or a culture somehow being weakened. It is never clear what is being threatened when, for example, White supremacists express fear of the demise of Western culture. I don't know what Western culture is, because it's very clear to me that my culture is not the same as the culture of other people in my street, zip code, city, country, or continent.

Nevertheless, the imagined end of this flabbily defined concept of Western culture is a permanent source of anxiety for White supremacists. They fantasize about a persecution of their people that will end in their extinction or an erosion of their rights in exchange for the same rights afforded to people of different heritage. When all you've ever known is privilege, equality feels like oppression. White nationalists marching in Charlottesville, Virginia, in 2017 felt compelled to carry tiki torches (a Polynesian technology) while chanting "Jews will not replace us!" The following day, amid civil unrest and violence between various factions of racists and anti-racist protestors, thirty-two-year-old Heather Heyer was murdered by a White supremacist. Her killer is now serving a four-hundred-year prison sentence.

The logic to arguments about who is entitled to be in a geographical region is often missing or at least ahistorical, as no people are ever static over long periods, and no power, culture, or nation has ever been anything near permanent. Nevertheless, our sense of family and ancestry is powerful, even if it is painfully restricted. For most people other than royalty, our family trees peter out beyond a few generations. The past is blurred by a paucity of records, by myth and lore.

But more than anything, our sense of our own ancestry is hamstrung by a failure to recognize a simple fact of biology: All humans have had two parents. When we look upward at our pedigrees, at best we can identify one, two, or a few lines into the past. At every branch on every family tree, we choose the one that bears fruit or we stop. We focus on notable, famous, or infamous people in our genealogical canopies, understandably because most people pass through history as shadows and dust, having lived normal lives that leave little or no trace, and any notoriety or fame is worth acknowledging. But in doing so, we ignore the vast majority of our ancestors who lived lives that vanished from history.

When it comes to my own family tree, we have no record of my Indian heritage, because records were lost after my ancestors were indentured to Guyana. On my father's side, we have traced a line of our ancestry to a great-great-great-great-great-grandmother of mine named Mary Huntley. Her wedding certificate says that she married Benjamin Handy in 1818 in Convent Garden in London. He was the sole proprietor of Handy's Travelling Circus and self-styled "The Greatest Horseman on Earth"—it describes Mary as "savage." She was the daughter of a Catawba tribesman, Neil Huntley, who moved from North America to join the circus for his own equestrian skills. This is undoubtedly a fabulous story, and it was a delightful surprise to discover an American Indian showman in my own family. However, Mary was one of sixty-four women to whom I am equally related in that tier of my family tree. The stories of the other sixty-three are lost in time.

In the study of genetics, we assume a generational time of twenty-five to thirty years, and in every generation back through time, the number of ancestors you have doubles. What this means is that over a five-hundred-year period, you have 1,048,576 ancestors. By a thousand years ago, you have 1,099,511,627,776—that is, over a trillion. This number is about ten times more people than have ever existed.* This paradox reveals quite

*The best estimates are that there have been around 107 billion members of our species, give or take a billion or so. This shuts down the popular myth that there are more people alive today than have ever lived. There

how incorrectly we think about our ancestry. The number of ancestors each of us has increases as we track back into the past, but the number of humans alive today is more than at any other time.

Both statements have to be true, though they appear contradictory. But the answer to this puzzle is obvious: Our family trees coalesce and collapse in on themselves as we go back in time. You certainly must have a trillion *positions* on your family tree, but the further you go back, the more frequently these positions will be occupied by the same individuals multiple times. It is quite possible that although I had sixty-four ancestral positions at the same tier as Mary Huntley, they may have been occupied by fewer than sixty-four women. Family trees coalesce with startling speed. The last common ancestors of all people with long-standing European ancestries lived only six hundred years ago—meaning that if we could draw a perfect complete family tree for every European, at least one branch on each tree would pass through a single person who lived around 1400 CE. This person would appear on all our family trees, as would all their ancestors. The fact that multiple positions are occupied by the same people indicates that the notion of a tree is again not the most accurate metaphor for describing

---

are more people alive today than on any other day in history. But the notion that the quick outnumber the dead is wrong by an army numbering about one hundred billion corpses.

genealogy: Trees only ever branch, but family trees contain loops. Your own pedigree rises from you like a tree, but sooner or later two of those branches will collide in a person from whom you are descended twice. These people sit atop genealogical loops.

Much is made of establishing celebrity in our tangled thickets, and royalty even more so. In 2016, on the popular UK television program *Who Do You Think You Are?*, British actor Danny Dyer discovered that he was twenty-two generations directly descended from the fourteenth-century British king Edward III. While few can establish this with genealogical records such as births, deaths, and marriages, according to my calculations,* the chances of anyone with long-standing British

*In order to calculate if anyone born in the 1970s was descended from Edward III, I counted how many descendants he had (until the number becomes fuzzy), and estimated the number of ancestors someone born in the 1970s would have at the same point in history. The question then becomes: "What is the probability that any of your ancestors at that time were in the proportion of the population who were direct descendants of Edward III?" He died in 1377, leaving thirteen children, six of whom had children themselves. I counted a total of 321 great-great-grandchildren, after which the tree becomes a bit too vague, but if we conservatively estimate that if each of these people on average had two children, then in the year 1600 Edward would have a total of around 20,544 descendants, which is large but far from impossible. The population of Britain at that time was around 4.2 million, which means that around 1 in 210 people was a direct descendant of Edward—approximately 0.5 percent of the population. There will have been around fifteen generations between 1600 and 1975, and the number of ancestors doubles each generation up a family tree, meaning that a person born in 1975 should have a maximum of 32,768 ancestors in 1600 (assuming

ancestry being similarly descended from Edward III is effectively 100 percent. It is true for Danny Dyer, it is true for the majority of British people, too, and it is true for millions of Americans whose ancestors decided that they didn't want to be ruled and taxed relentlessly by a greedy monarchy five thousand miles away.

Go back a few centuries further and we reach a mathematical certainty referred to as the genetic isopoint. This is the time in history when the entire population is the ancestor of the entire contemporary population today. For the people of Europe, the isopoint occurs in the tenth century. In other words, if you were alive in the tenth century in Europe, and you have European descendants alive today, then you are the ancestor of *all* Europeans alive today (we estimate that up to 80 percent of the population of tenth-century Europe has living descendants). Another way to think of it is like this: One branch of a family tree of two first cousins crosses in a

---

full outbreeding, which is very unlikely, but makes little numerical difference to the calculation). Therefore, each one of your 32,768 ancestors in the year 1600 has a 0.5 percent chance of being a direct descendant of Edward III. If you reverse the question, and ask, "What are the chances that none of your 32,768 ancestors in 1600 is in that 0.5 percent?" the calculation becomes: $0.995 \times 10^{32,768} = 4.64 \times 10^{-72}$. To get the answer to the original question, we subtract that from 1, which comes to: 0.999999999999 (etc.). If you have any broadly British ancestral lineage, you are descended from Edward III, and all of his regal ancestors, too, including William the Conqueror, Æthelred the Unready, Alfred the Great, and, in fact, literally every tenth-century European ruler and peasant.

shared grandparent; one branch of all European family trees cross through one individual in 1400 CE; at the isopoint, all branches of all family trees cross through all people for that population.

I am well aware, having said these facts to students and public audiences hundreds of times, that this is a brain-scrambling concept, because it is so far from our casual assumptions and thoughts about ancestry, family trees, and identity. It certainly doesn't sound right, and is further confounded as a concept by the calculations of the *global* isopoint—the year in which the population of Earth was made up of the ancestors of everyone living today. This, astonishingly, comes out to around 3,400 years ago. Everyone alive today is descended from all of the global population in the fourteenth century BCE.

Irrespective of how plausible that sounds, or how contrary it seems to our own experiences of family and family trees, it is true—the isopoint is a mathematical and genetic certainty. It is likely that the proportion of a person's ancestors at the isopoint is not equally distributed around the world: A Chinese woman or man will have far fewer southern African ancestors than East Asian, and vice versa. But they will have some, and each of those ancestors has an equal relationship with their living descendants regardless of where on Earth they lived and died.

We think of certain areas, lands, or people being isolated either physically or culturally, and these boundaries are insurmountable. But that is neither what history nor genetics tells us. No nation is static, no people are pure. The global isopoint might have been much earlier if it weren't for the expansion of Europeans. The first people of the Americas were isolated in that continent from around twenty thousand years ago, when they had crossed from Siberia on dry land exposed by an ice age that had sucked water into its glaciers and lowered the seas. But when the thaw came, the people who had moved into what is now Alaska were cut off from the rest of the world for more than fifteen thousand years.

There were a handful of migrations from Asia in the last 4,500 years, including by ancestors of Inuit today. One thousand years ago, the Vikings, under the leadership of the Icelander Leif Ericson, had a fleeting three-year sojourn to the American continent, in places now known as Labrador, Baffin Island, and Newfoundland in Canada, but left no lasting legacy or genetic trace—after an argument about a bull, those fierce warriors were chased away by the indigenous people they called Skraeling. But when Columbus and his men invaded the Caribbean in 1492, the rape of indigenous Taino women began immediately, and European ancestry was introduced into the people of the Americas. In only a few generations, this admixture percolated in all directions, and those genetic signatures are found in

Americans north and south, regardless of how isolated you might imagine these tribes may have been.

These ideas of how ancestry and family trees actually work make pure mockery of the concept of racial purity. Sometimes people write to me stating that they can trace their ancestry through centuries, and it locates to one geographical area. This is often held to be a badge of honor, a lineage of centuries that bestows some sense of personal or tribal identity upon them. A friend told me of the family story that they were descended from Niall of the Nine Hostages, an Irish king and ancestor of the medieval Uí Néill dynasties. Niall was a fifth-century ruler, if he existed at all, and so if my friend's statement is correct, it is also true for literally every European too. One proud and stubborn Irishman once told me how his ancestors were *all* from a small area in Ireland that they could trace back for a thousand years, and refused to accept that many of his ancestors must have come from all over the place. He wasn't a racist, but if he was right, then he would be dangerously inbred.

Some people assert racial purity for similar reasons. It is true that for many people a large proportion of their ancestors will be from one area over a time span of decades or even a couple of centuries. Despite the concept of the isopoint, we don't randomly mate in a globally distributed full shuffle. In my stepmother's family, we can trace back eight generations in a single graveyard in the county of Essex in southeast England. Did all of her 256 ancestors at that tier come from Toppesfield or neighboring villages? Of course not. The slightest movement of people changes our family trees, introducing new people and new lines, and the trees are far from arboreal and much more tangled. Only one of my ancestral lines is American Indian and that is a fun story to tell. Does it mean I am American Indian? It certainly does not.

People have moved around the world throughout history and had sex wherever and whenever they could. Sometimes these are big moves in short times. More often people are largely static over a few generations, and that can feel like a geographical and cultural anchor. Nevertheless, every Nazi has Jewish ancestors. Every White supremacist has Middle Eastern ancestors. Every racist has African, Indian, Chinese, American Indian, and Aboriginal Australian ancestors, as does everyone else, and not just in the sense that humankind is an African species in deep prehistory, but at a minimum from classical times, and probably much more recently. Racial purity is a pure fantasy. For humans, there are no purebloods, only mongrels enriched by the blood of multitudes.

We can now pick out some of the threads of deep ancestry with DNA, and this helps us understand broad sweeps of human migration and, to a lesser but still exciting extent, the tighter, smaller movements of people in history. We can see these patterns in the genetics of living people, and while the stories of early Europeans are now settling into robust narratives, for much of the world there is still a great deal to be discovered.

As discussed in part 1, Africa is not yet well represented in terms of understanding the genetic histories of its people. We have seen that there is much more genetic diversity within Africa than in the rest of the world put together, which means that people within Africa are on average more different from each other than anyone else on Earth is from each other. This is a reflection of the Out of Africa population being small, and therefore not representative of the people whence they came. Only a small proportion of people left Africa to become the pool from whom the rest of the world would be drawn.

A much larger population did not. Africa is a huge continent, and for seventy thousand years people have migrated and swapped genes in every direction *within* that continent. As I mentioned earlier, there has also been some European and Middle Eastern backflow, where, in the last few thousand years, people have meandered back into Africa and spread some of their genes into African genomes.

As a result, African genomics is complexly entwined, and has not been studied in as great detail as European DNA. We are only just beginning to untangle the movement of people within Africa, from tribes to city-states, within and between countries. The power of genetics as a historical source is only starting to be applied in the cradle of humankind, and some of the tales it tells are inspirational. The Kuba Kingdom was a territory in what is now the Democratic Republic of Congo, and while the Kuba existed in and around that area since the sixteenth century, there was a period of dramatic growth and prosperity that occurred independently and before Belgian colonization. The growth, according to oral histories, was facilitated by a charismatic king called Shyaam, who united local tribes, who migrated into what became a centralized city-state, with many of the characteristics of modern political systems that were largely absent at that time: a capital, an oral constitution, a tiered legal system, trial by jury, taxation, and a police force. Following colonization, the kingdom was weakened, but it still exists within the Democratic Republic of Congo, and many people identify as Kuba. As is now possible, the story of the Kuba is one that can be tested using DNA, and Lucy van Dorp from University College London led a team that did exactly that. By sampling the DNA of 101 people with Kuba ancestry and comparing it to several hundred people from other local populations, they showed that Kuba people had a far greater mix of DNA from around the region, indicating that the lore about the fusion of diverse and disparate groups via immigration and integration is indeed true.

Much of the discussion in this book concerns the misalignment of genetics with race from the perspective of racism by European colonizers on the rest of the world. Racism with a pseudoscientific basis is not unique to Europeans subjugating others. It is worth noting that there is plenty of racism within Africa, and around the world, that is similarly impossible to justify from a biological perspective.

In 1990, during the Rwandan Civil War, the population known as Tutsi was massacred by the insurgent Hutu people. Estimates vary, but some plausibly suggest that up to a million people were murdered, some 70 percent of the Tutsi population, maybe 10 percent of the total Rwandan population—a literal decimation in one hundred days.

This was a race war. Much of the basis for the animosity and subsequent genocide was the belief that the Tutsi and the Hutu are genetically distinct, and the genesis of this belief is directly drawn from colonial rule. During the years of German occupation in the nineteenth century, tribal relations were kept largely positive, such that the colonizers could use the industry of locals to maximize their extraction of valuable crops and commodities. German colonizers thought that Tutsis were superior to Hutu, and that this might have been because of Hamite ancestry, that is, a Caucasian race and linguistic group invented in the early nineteenth century that supposedly branched from Middle Eastern populations. The origin of the term is that these were a people descended from Noah's son Ham, who were "cursed with blackness" according to a passage in the Talmud. This ancestry, the colonizers thought, meant that the Tutsis were superior to other Africans.

When Belgium took over in the early twentieth century, they sowed and cultivated seeds of racial disharmony. Adopting racialized pseudoscience derived from the contemporary eugenics movement, Belgian officials asserted that Tutsis had larger brains, lighter skin color, and a higher frequency of milk drinkers, and concluded that they had European ancestry and, like the Germans before them, that they were therefore superior to Hutu and other ethnic groups.

When ethnic identification cards were introduced in 1933, the racialization of these two groups was formalized and, crucially, adopted by both Tutsi and Hutu. Conflict between them was continuous during the twentieth century, and as the Belgian colonizers left in the late 1950s, the Tutsi monarchy was replaced during a violent Hutu revolution.

The infamous genocide that began in 1994 during the Rwandan Civil War was instigated by the Hutu government; hundreds of thousands were murdered, and rape was weaponized on an industrial scale. These decades of conflict, murder, and genocide were predicated on claims of racial distinction and purity, all built

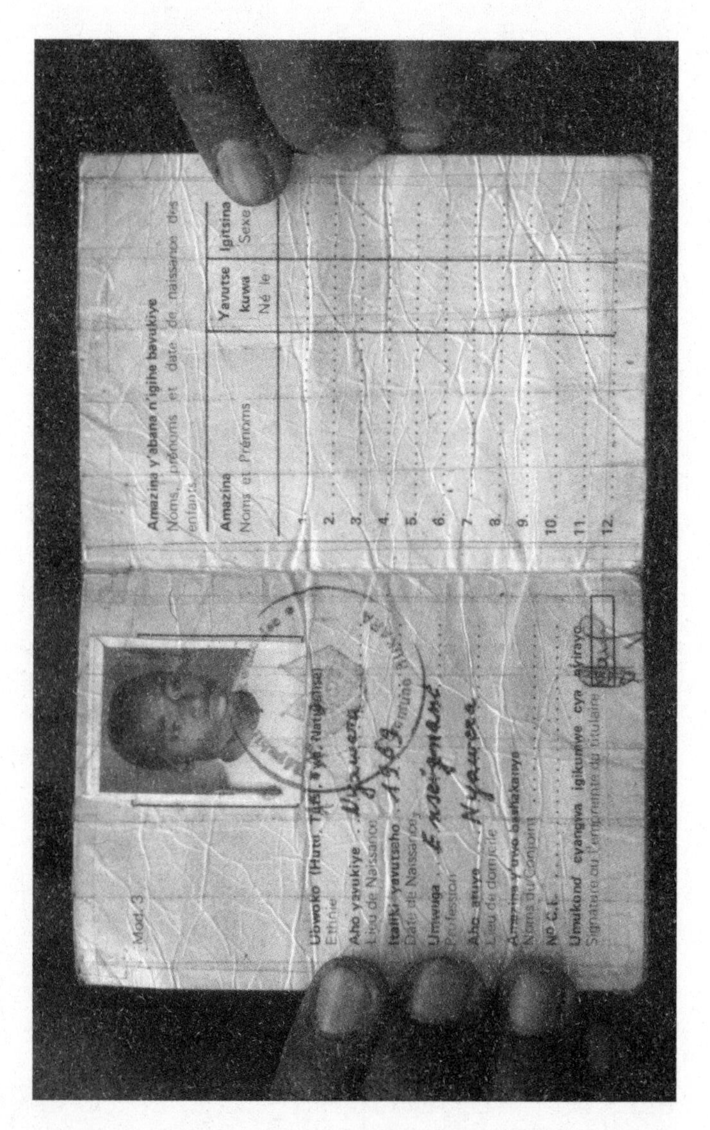

A Rwandan ethnic identification card: Note the ethnicity line just below the photo; the only label not crossed out is "Hutu."

upon pseudoscience. The anthropological, anthropometric (measurement of the body), and phrenological (measurement of the skull) bases for these claims were all bogus, and derived from centuries of European scientific racism passed on to groups that became hostile rivals.

The genetics of the people of Rwanda is complex, as it is in so much of Africa, and while there are some measurable genetic dissimilarities that indicate populations have had different ancestral pathways, they are hugely overlapping. Social and cultural practices may well be different—Tutsi were traditionally more pastoralist, which may account for higher levels of lactase persistence and thus milk drinking—but that's not much of a basis for ethnic cleansing and genocide. For the most part, there has been plenty of admixture between the Tutsi and Hutu, and like in so many civil wars, the biological difference was negligible. The grimmest of ironies emerges from this horrifying tale: Up to ten thousand war babies were born as a result of the perpetration of rape as a weapon of war. These children carry the genes of Hutu and Tutsi. The result was not ethnic cleansing but ethnic admixture.

A slightly different, but no less pernicious, version of racial purity fixates upon displacement of people. In Britain, while there is contemporary angst about immigrants and refugees, those on the far right have long expressed anger in the form of slogans such as "England for the English," or some make an argument based on protecting citizenship for the indigenous Brits—something that, to the best of my understanding, is not under threat. "Go back to where you come from," someone told me on Twitter last year, and I did indeed drive up the highway to Ipswich to visit my folks for the weekend. In July 2019, President Trump stated that four elected US congresswomen "originally came from countries whose governments are a complete and total catastrophe" and that if they didn't like it in the US, they should go back. Three of them were born in the United States and one, Ilhan Omar, is a Somali-born American citizen. Donald Trump's paternal grandparents were German immigrants to the US, and his mother Scottish born, his first wife born in Moravia, his third in Slovenia. It is never clear *when* the commonly considered benchmark for indigeneity is.

In a trivial way, it's problematic for people such as me who have recent ancestry from abroad, or for British Blacks, or South Asians descended from postwar immigration, and I suspect much of the ire of racists is directed at us. But Britain has been steadily and continually invaded throughout its history, and has become home to migrants since it became an island around 7,500 years ago. In 1066, the French came and enacted a hostile takeover with an arrow to the king's eye. Before that, England was invaded by Vikings, aggressively, and before that there was continual movement of people from the continent, Angles, Saxons, Huns, Alans, and dozens of other small tribes and clans. Before that, the Romans ruled, at least as far as Hadrian's Wall to the north, yet many of the Roman army's conscripts were not from Rome but from all over that expansive intercontinental empire and beyond, and their ranks included Gauls, Mediterraneans, and sub-Saharan Africans.

All countries are different, but all have enjoyed and endured constant and continual immigration and invasion, and the mixing of people that had previously been separate, because that is what humans do. Britain's deep history is well studied, and makes a robust demonstration of this picture of history. Around 4,500 years ago, my homeland was populated primarily with farmers whose people had migrated from Europe, over Doggerland, the continuous terrain that is now the North Sea between East Anglia and the Netherlands. These immigrants were the people who built megalithic oddities such as Stonehenge. On the basis of DNA evidence, we think they may have been olive-dark skinned, like southern Mediterranean people today, with dark hair and brown eyes (as best we can predict, given the caveats we explored in part 1). Over in continental Europe, a new culture was emerging, which spread widely in a short period of time. We call the people who exhibited it the Beaker folk, after the characteristically shaped pottery jars that are found in burials and other sites from this time. We don't know if there was a central origin of this type of material culture, but soon it was all over Europe. The culture and the people who came with it arrived in Britain about 4,400 years ago, and according to the DNA retrieved from bones in these lands, within a few centuries they had replaced almost the entire population, a turnover of genetic identity greater than 90 percent. Their dominance did not last long. We don't know how or why, whether it was violence, disease, or something else, but after only a few centuries, they were all gone, and Iberian farmers with their distinctive bell-shaped pottery and cinerary urns had become British.

Before the people who built Stonehenge, there were others, hunter-gatherers who had been there for a few thousand years and were darker skinned. Cheddar Man, from part 1, who died ten thousand years ago, was one of them. And before them, well, it gets a bit fuzzy. In the parish of Boxgrove in Sussex, in South East England, we have bones from another species of human, probably *Homo heidelbergensis.* It was a tall woman or man, from about half a million years ago, who hunted rhinos and bears whose bones are also found nearby. But the earliest evidence of British people is in the crumbling coastline of Happisburgh (pronounced *Haze-bruh*) in the eastern county of Norfolk, where size-nine footprints were set in soft stone nine hundred thousand years ago and were revealed only when the tide was low.

This grand picture is not fundamentally different for any nation. Only the timings and the details change: New Zealand was humanless until around the eleventh or twelfth century, the Americas received their first peoples only some twenty thousand years ago. The only true indigenous Brits lived almost a million years ago, and we are not sure what species they were. So, when racists say Britain is for the British, or when they talk about indigenous people, I do not know whom they mean or, more specifically, *when* they mean. I suspect that they don't either.

Geological history and the history of humans pay little mind to the transience of borders and governments. The extensive British colonial past means that the evolution of citizenship is complicated by a history of empire. But if you are a British citizen, you are entitled to a British passport, which legally, technically, and actually makes you British. This is a nonnegotiable fact. The presentation of arguments based on who are "real Britons," or the "indigenous people of Britain," is an ahistorical, nonscientific smoke screen to hide racism.

Not all countries are the same though. "First people" is not a simple concept, because everywhere habitable on Earth has hosted people for almost a thousand years, New Zealand being the last significant landmass that humans reached. In a legitimate sense, the Māori are an indigenous people, as they were the first humans to set foot on Aotearoa, as they call those islands. By the time they arrived in the eleventh or twelfth century, Britain had been invaded aggressively for the last time. Vikings were the first men to set foot on Iceland (with the possible exception of an Irish monk or two—though, being pious and chaste, they didn't leave any descendants). The Vikings were Norwegian and Danish men, who had picked up Scottish, Faroese, and Irish women on their voyage west. Were the indigenous people of the Americas a pure race by the time Columbus invaded? No, because they spent more than twenty thousand years migrating *within* a continent that spans almost the full longitude of Earth.

There is another, further confounding point when it comes to DNA. Basic biology tells us that we inherit half of our genome from our mothers and half from our fathers. This is a truth universally acknowledged for all humans through time:* A new whole genome is forged at the conception of a child. But the process of genetic shuffling that occurs in the formation of sperm and egg guarantees that each one of those two cells is unique and carries a unique half genome (therefore a unique half is lost in subsequent generations, should that sperm or egg be successful). Which means that not the same half gets transferred each generation. Over the generations,

---

*Though in the last few years, a few children have received mitochondrial genomes from a third genetic donor to cure diseases. "Three-parent children" is how the press is fond of describing them, but the truth is that it's such a tiny amount of DNA, that moniker is not really warranted.

descendants begin to shed the DNA of their actual ancestors. The amount that vanishes is cumulatively huge: You carry DNA from only half of your ancestors eleven generations back. Genealogy and genetic genealogy are not perfectly matched, and progressively grow apart as we go back in time. It is therefore possible that you are genetically unrelated to people from whom you are actually descended as recently as the middle of the eighteenth century. This is a point that further undermines the appropriation of genetics as a means of asserting membership of a tribe, race, or other identity.

As we've seen, sequencing of DNA became so cheap and quick a few years ago that companies sprang up that would take your genes, typically from a saliva sample, and read specific parts of your DNA to make some predictions or claims about any number of the personal variants that you harbor. Some of these companies focus on health or diet, sporting ability, or even more preposterous claims such as wine-tasting preference or compatibility with a potential spouse. A few companies have become vast industries, and these are rooted in the business of genetic genealogy. Some companies have come and gone, having made outlandish claims about membership of ahistorical tribes, wandering nomads, or romanticized potters. The natural selection of market forces has winnowed away some of those businesses all the way to extinction, but of those that remain, the giants are 23andMe and AncestryDNA. These two now effectively have possession of the genomes of something like twenty-six million customers who have paid to give up their spit and therefore DNA in exchange for some information about their genetic heritage.

The marketing for these services is persuasive and alluring. They typically appeal to our narcissism, and to our sense of curiosity and belonging, with messages such as "Find your roots," or talk of exotic or unknown ancestors. 23andMe capitalized on the 2018 men's soccer World Cup finals with advertisements suggesting you should "Root for your roots: Be the ultimate soccer fan by supporting the countries that reflect your unique DNA."

I guess the marketing department hadn't met many soccer fans.

Customers on the Ancestry website tell testimonial tales of personal discovery and identity: "I started considering how much of my identity was defined by my family history," says Mark. "When I was young, I always thought I was 100 percent British. My dad was born in Edgware and my mum in Hampshire." But according to the ad, he discovers that he has Russian, German, and Greek great-grandparents, and DNA tests tell Mark that he is "only 40 percent British, 25 percent German, and 35 percent Greek."

These are not racist sentiments of course. Quite the opposite, they are promoting the notion that we *are* a happy mix, with ancestry from all sorts of places that might be new to us because of lost or unknown family narratives.

That doesn't make them scientifically robust though. What these services are actually doing is comparing your DNA to databases of other customers—that is, other living people—and charting where on Earth they live today. The maps that you receive after a few weeks show your similarity to living populations, and from that you are to infer ancestral roots. This is not incorrect, as it reveals populations who have made a genetic contribution to your genome. It reveals a probability of a proportion of ancestry. The kits can be very good at identifying very close family members, and there are a few reported cases of the discovery of lost siblings, cousins, or unknown parents from adoptions. But for the vast majority of paying customers, the results are broad and bland.

The fudging of the data to say that you are "40 percent British, 25 percent German, and 35 percent Greek" or some other combination is confusing to me, and it doesn't indicate the number or relation of the ancestors who have longer-standing Greek heritage. A more accurate result would say, "Despite the fact that your genome has significant genetic contribution from people who have recent geographical association with the modern nation-states of Germany and Greece, though we can't be sure which of your ancestors these were, your family tree spreads all over Europe and, to a lesser but still significant extent, indeed the world. However, you remain 100 percent British because that is how citizenship legally is determined. Genetics won't change that." Admittedly that is an unpithy marketing campaign, and perhaps a less desirable product to give as a Christmas present.

These types of services and results also reinforce a long-standing belief in a kind of essentialism that comes with nationhood. This belief is a characteristic that besets popular understanding of inheritance and genetics. Genes play a significant role in all our biology, including our behavior. Aside from the fact that we can measure that contribution in contemporary populations, and that it is heritable, we don't necessarily understand how it works, and certainly can say very little about the stability of such genetic contributions to traits over generational time. Does "35 percent Greek" mean anything about your character or behavior? Frequently and casually, people tell me that their unforeseen ancestry somehow accounts for their personalities based on national stereotypes, and these are invariably positive or attractive traits: fiery Spaniards, for instance, or methodical Germans, passionate French, hardy Scots. No one has ever said to me that it is their ancestry that accounts for the fact that they are weak-willed, scared of spiders, or lily-livered lickspittles. I'm sure that there may be some shard of truth in the idea of national characteristics, as people who live together with the same cultural influences over generations can and will behave more similarly than with others. That these might be encoded genetically, be stable through time, and explain personal behavior is doubtful.

There's yet another factor that undermines the informativeness of these types of ancestry tests. The results are dependent on comparison within a database comprising the DNA of other paying customers, and not a random or general population. Instead, your results are most likely to be determined by people who are similar to each other: For socioeconomic reasons, those paying customers tend to be relatively wealthy Europeans or European-descended North Americans. The resolution of the resulting data is extremely high, for example, for my 23andMe European genome, itemizing proportions of DNA that are most similar to populations in regions of Scandinavia and France and within Great Britain. Half of my genome is from my Indian heritage, but in the same results, those 1.3 billion people are currently represented by a single uniform block devoid of any structure or detail, simply because relatively few Indians or people of Indian descent have bought these kits and handed their DNA over to the database.

In the US, this problem is heightened, as you might expect in a nation with such a recent and peculiar history. Around one-eighth of the population is Black, descended from people enslaved largely from West Africa.* The actual birthplace or citizenship of their ancestors is almost always entirely unknown. The indigenous people of the Americas, specifically American Indians, make up about 2 percent of the total population of the US. Regardless of current laws outlawing racist practices and current levels of racism in the populace, both groups have been subject to recent historical, governmentally sanctioned racist policies, the civil rights revolutions having happened only in the 1960s and forced sterilization of American Indians occurring as recently as the 1970s.

Both groups are likely to be in lower socioeconomic demographics and have similarly low uptake of genetic ancestry kits. Having said that, some companies have focused their products on these demographics accordingly, with no less vapid conclusions. African Ancestry is one such company, and on its website says that unlike the competition, they can "identify an African country of origin" and "specify an African ethnic group."

Both of these statements, in my opinion, are scientifically questionable. Ethnic groups within Africa are often more cultural than genetic, and don't correlate particularly precisely with clustering of populations when sampling genomes. In some scientific analyses of genomes

---

*In reality, many Americans of European descent carry African ancestry, and many African Americans have a significant proportion of European ancestry.

from around Africa, the authors shy away from the resolution of specific countries but can identify, for example, a genetic signature that covers "western Bantu-speaking ancestry"—the Bantu being a hugely diverse grouping that spans the width of the continent, and hundreds of millions of people organized loosely into hundreds of tribes. Furthermore, the most recent genetic studies within sub-Saharan Africa indicate a profoundly complex genetic history of Africa. The flow of those genes into the Americas is equally scrutable.

We've already touched upon the complexity of the genetic structure of people from the African continent. Starting with that baseline, the movement of people to the Americas during the Atlantic slave trade era complicates things even further. Estimates vary, but historians generally consider that between the sixteenth and nineteenth centuries some twelve million people were taken from coastal countries including Senegal, Sierra Leone, Angola, and Congo and brought to the Americas, north and south. This book is not a history of slavery, but there are some pertinent points to make with regard to attempting to understand one's ancestry via genetic testing. The first indentured Africans arrived in the then English colonies at Jamestown four hundred years ago, though Africans were present in the Caribbean and northern America for a century longer. Slavery was instituted inconsistently from the seventeenth century onward under colonial law, notably under the principle of *partus sequitur ventrem*, meaning that a child born in the English colonies would inherit the legal status of their mother—a daughter or son of an enslaved woman would themselves be born enslaved. Part of the foundation of this law was in response to the pivotal case of Elizabeth Key in 1656. She was the daughter of an African woman and an Englishman, and is recorded in court records as "molleto" (better known as mulatto, meaning mixed race). Key successfully sued for freedom for herself and her son John, on the grounds that she was baptized a Christian (who, at the time, were not allowed to remain permanently enslaved) and the daughter of an Englishman. Her son also had an English father, in this case her English husband and attorney William Grinstead; at that time, the status of the child was determined by that of the father. Key remarried after his death and John was a freeman. As this was a family of historical significance, their descendants are well documented: Many have the surnames Grinstead, Grimsted, or Greenstead, and they include the actor Johnny Depp.

The law of partus was introduced in 1662 by the General Assembly of Virginia to excuse paternal responsibility for children White men fathered with enslaved women, a concept known as hypodescent, where social status of children of mixed ancestry is allocated by the dominant group to the subordinate group. White men fathering children with enslaved women was a common occurrence, most famously in the post-revolution era in the case of President Thomas Jefferson, who is alleged to have fathered six children with Sally Hemings, herself the daughter of a partnership between an African American and an Englishman. Under Virginia law, Jefferson's children were legally White owing to the proportion of their English ancestry by family tree, but born enslaved because of partus. Many of their descendants are also known today.

Though importation of chattel slaves was officially outlawed in the US in 1808, the slave trade itself continued within America until President Lincoln signed the Emancipation Proclamation fifty-five years later, and 3.5 million enslaved Americans were freed.

This very superficial description of American life over four centuries has profound implications for understanding ancestry within the US today. The US population at the end of the Atlantic slave trade was around seven million, and twenty-three million by the end of slavery. Immigration accelerated over the next century, and the population expanded to the 325 million Americans today, including some from Africa, but the majority from European countries. Today, the African American population of the US is around forty-two million. Consider that there was continual interbreeding within enslaved peoples, and between the enslaved and their owners, and with the same application of general rules about generation time in humans, it is virtually inconceivable that a genetic test can establish an African country of origin from transatlantic slavery. As with everywhere on Earth, an African American today will have more than one thousand ancestors in the eighteenth century. They cannot have all come from one tribe or country.

Millions of Africans were transported, and millions died en route, from disease or by jumping from the ships because they knew death was better than bondage. The survivors of the journeys were not kept separate by country of origin, nor could they possibly have been when being traded like cattle in plantations around the Americas. Perhaps as databases grow and work continues with ever more fine-scale analyses, it might be possible for DNA to identify that some ancestors were from particular regions, or even particular tribes as they stand today. But as everywhere, even with this grotesque history in place, everyone has two parents, four grandparents, eight great-grandparents, and so on,* and with notable levels of admixture within Africa, the purity of the genetic signals that might reveal something as precise as country will be blurred. The messiness of human movement and the desire to reproduce, be it via consensual partnership or acts of cruelty and wickedness, render the concept of a singular geographical origin nonsensical.

---

*It is worth noting that the offspring of the two first cousins who marry, for example, have only six great-grandparents, but they still have eight great-grandparental *positions*. It's just that two of those grandparents will each be in two of those positions.

The desire to know something about one's ancestry is powerful, and in the case of African Americans, empathy is important. The Atlantic slave trade was a pernicious nadir of man's capacity for cruelty to fellow humans. Ancestral homes were destroyed, tribes annihilated, countries decimated. Millions died on the ships to which they were chained. For a people—multiple peoples in fact—to be so uprooted from their past, for it to be a blank sheet, any information might have some validity or offer some empowerment or comfort. Nevertheless, the commercial genetic tests remain scientifically unconvincing.

For American Indians, the story is different, but the outcomes similar. The oppression and persecution of the diverse indigenous peoples of the Americas began in 1492 and lasted for centuries. During that time, tribes were forcibly relocated, women raped and murdered. The Trail of Tears is perhaps the best-known forced migration in US history. In 1830, President Andrew Jackson ratified the Indian Removal Act, which, while ostensibly merely giving the federal government the right to open voluntary relocation negotiations with Cherokee tribes, had the intended effect of facilitating the forced relocation of more than sixteen thousand Cherokees, following the discovery of gold in Cherokee territories.

Thousands died during this forced exodus. That these types of genocidal policies existed reflects the inherent racism of successive governments over decades, but it also describes a population history that is very abnormal, on top of twenty thousand years of migration and expected levels of admixture within pre-invasion America. The existence of few written records of ancestry within American Indian populations, coupled with a paucity of genetic samples, means the current status of our understanding of American Indian genomes is relatively poor. We know that there was gene flow between tribes before colonization and after. We know that forced migration means that membership of tribes has included some flux due to relocation and the sense that tribes are very much linked to the land they inhabit. There are several ways in which tribal status is assigned, primarily from a concept called blood quantum—an invention of European Americans in the nineteenth century—which concerns how many of your ancestors are already in a tribe. Aside from cases of challenged paternity, DNA cannot be used meaningfully.

That hasn't stopped the emergence of genetic genealogy companies that sell products specifically claiming they can assign exactly that. According to Accu-Metrics, there are "562 recognized tribes in the U.S.A., plus at least 50 others in Canada," and for $125 they "*can determine if you belong to one of these groups.*" DNA Consultants sells a Cherokee test for $139, and for an extra twenty-five dollars you can get a certificate. These products are pseudoscience, genetic astrology, in my view. Given the paucity in the current databases of American Indian DNA, I am of the opinion that it is currently impossible to ascribe tribal status using DNA, and given the population history of indigenous tribes, I believe that it will never be possible.

We now are confident that long-range migration and relentless exchange of genetic material has been a ubiquitous feature of human history and that, as a result, current population structure is not necessarily a good proxy for the geographic locations of ancestral populations. Every nation on Earth is unique, and all are the same. There is no such thing as racial purity, and genetics has made of a mockery of such claims. Populations around the world do have genetic signatures that reveal the current and to some extent the historical structure of the people who bear them. But these correspond poorly with any concept of race, or even country.

The discussion so far has focused on attempts to spot cultural identities using genetics, which, at best, is a struggle. The same naturally applies to people of European descent who claim racial purity and therefore racial superiority. Racism as a concept has multiple definitions, but all are quintessentially comparative. However a group is racially defined, the implication is of behaviors or traits that can be used to rank groups.

The Venn diagram of people who describe themselves as White nationalists, White supremacists, and neo-Nazis is close to being a single circle, though they all claim subtle differences. As in the early days of scientific racism, almost all self-define as superior to other races.* Since the advent of the Internet, which is pretty much synchronized with the genomics revolution, racist websites have existed. Possibly the most widely known is Stormfront, which describes its membership as "racial realists" and White nationalists, but there are plenty of others, including the influential chat forums on websites such as 4Chan and 8Chan. Stormfront also states up front in its introductory pages its specific interest in genetics:

> The problem with humanity is not so much one of ideology—this or that religious, political, social, or economic system—but rather one of blood. That is, that a great deal (possibly 90 percent or more) of a person's intelligence and character is determined by their DNA, which determines the structure of their brain before they are born. This is why Blacks, as a group, do the things they do.

One of the specific aims of many White nationalist groups is to establish some kind of White ethnostate, and since the rise of commercial personal genomics tests, these sites are filled with racists obsessed with population genetics. It isn't quite clear how racial purity would be established, but services that genealogy hobbyists

*The American Nazi Party asserts that they are not White supremacists, merely separatists. They also specifically want to ban modern art and rap music. So it goes.

use in attempts to trace their ancestry are also popular with those who use them to demonstrate some concept of White purity. These sites are crammed with comments showing off their test results, as long as they indicate northern European ancestries. Prominent White nationalist Richard Spencer posted his 23andMe results to Twitter in 2017, which showed 99.4 percent European and no Ashkenazi Jew. As is possible on the 23andMe website, he also allowed the full result to be seen by all, which reveals that he has North African and Mongolian ancestors as recently as the nineteenth century. Funnily enough, Spencer has yet to comment on this.

In these forums, there is also a noticeable interest in technical scientific papers, a voluminous discussion of studies that would not normally make it far past coffee rooms in academic institutions. Levels of comprehension vary enormously, but some contributors do at least have an understanding of basic genetics, and go to some lengths to explain it to others in these forums. They inevitably lack nuance or draw different conclusions from those in the paper, or simply deny that these studies are correct. A further activity in these cesspools is to take figures from academic papers and relabel them to create memes to be distributed on other social media platforms, such as Facebook and Twitter. Any geneticist who has shared results, data, or opinions on these types of scientific papers on social media knows that the deluge of racist replies can be overwhelming. Responses appear semi-coordinated, sometimes with the same wording or memes being used repeatedly. This can be a surprise to scientists who perhaps have not been exposed to it before, or were unaware that their work was being discussed in depth in racist forums.

Anecdotally, there are also accusations that geneticists in public life make statements about the scientific invalidity of race in relation to specific abilities (notably cognitive abilities and intelligence, which we will explore at length in part 4), but within the safe spaces of the academy, we actually think and say something different. Such accusations have been made in articles in major newspapers, and to me personally by prominent media figures with large social media platforms. This, it is almost embarrassing to have to say, is batshit-crazy conspiracy garbage. It is comically insulting to thousands of scientists, whose lives are dedicated to pursuing objective truths about people and nature, and it is devoid of any evidence in support. The idea that we are hiding some truth from the public for political reasons is absurd. As with equally mad antiscientific ideas such as creationism, if I could demonstrate that Darwin was wrong or that race is a scientifically valid and useful description of human variation, I would be the most famous biologist in history, and the riches that would follow would surely be magnificent.

As Jonathan Swift said in 1721: "Reasoning will never make a Man correct an ill Opinion, which by Reasoning he never acquired."

Arguing with racists with conspiracy mindsets about science is a fairly fruitless endeavor, and exhausting. To be so locked in and fixated on a facile idea is an entrenched stance. While haunting these racist forums, particularly ones that focus on commercial genetic ancestry tests, one sees the occasional discussion of results that appear to reveal previously unknown heritage from people whom White supremacists despise. I will not hide my bitter enjoyment of these tiny shafts of light in otherwise dark pits. In 2017, a study accounted for this exact phenomenon. What happens when you have committed to a racist ideology, only to discover you have recent ancestors from populations that you hate?

More than three thousand comments on Stormfront were analyzed by sociologists Aaron Panofsky and Joan Donovan. Stormfront claims several hundred thousand users, and is the biggest and longest-standing racist Internet forum (they make internal distinctions between White nationalists and supremacists, which are not particularly relevant here). Comments that described results that confirmed the users' beliefs in their racial purity were typically expressed in terms of relief or pleasure, with expressions such as "pure blood" or "100 percent White."

In discussions about genetic genealogy concerning discovery of non-European or non-White ancestry, various strategies were employed to question or parse the results—some sophisticated, others as dumb as bricks. Of the less sophisticated responses, paranoia or conspiracy abounded: *The companies are owned by Jews*, or *they are part of a plot to sow doubt about racial purity*. This is par for the course on Stormfront, which states in its introduction:

> The Jews have been working together behind the scenes to gain control of all the TV stations, schools, newspapers, radio stations, governments, movie studios, banks, etc. . . . The origin of the problem with the Jews is, once again, in the blood. As a group, a distinctive *race*, they suffer from psychopathy—a mental disorder whose main symptom is the ability to lie like there is no tomorrow.

Other responses involved attempts to dismantle or dismiss the results of ancestry tests. Fractionally more sophisticated than the Jewish conspiracies were the assertions that the data itself was flawed for that particular testing company, and the recommendation that the user should try a different one. Some dismissed low levels of non-European admixture as noise or not noteworthy, though the threshold of non-significance was arbitrary and varied enormously. Some conflated the percentages with assumptions about what proportion of genealogical ancestry that bestowed, rather like the rules established in the slavery-era United States to determine whether a child was born White or some other designation such as "mulatto" or "quadroon."

Panofsky and Donovan also document responses to users revealing non-European or Jewish heritage (Stormfront requires registered users to have no Jewish ancestry—something that is pretty much impossible for all Europeans). Replies included the sympathetic: "I wouldn't worry about it. When you look in the mirror, do you see a jew [*sic*]? If not, you're good"; and extreme hostility, suggesting they should be barred from the site or should commit suicide. One user revealed 61 percent European DNA, to which another replied: "I've prepared you a drink. It's 61 percent pure water. The rest is potassium cyanide. I assume you have no objections to drinking it. (You might need to stir it first since anyone can see at a glance that it isn't pure water.) Cyanide isn't water, and YOU are not White."

White purity is the key idea within White supremacy. Whiteness is perceived as being superior to other pigments, not least because of an interpretation of history that puts Europeans as dominant over other countries via conquest and empire, as well as somehow bestowing characteristics such as being inventive and wealth creators. These attitudes are remarkably similar to those expressed throughout the history of scientific racism by

Kant, Voltaire, and many others from the seventeenth through to the twentieth century. Admixture with people other than Whites is an act of dilution away from the purity of White ancestry, and therefore undermines justification for a White ethnostate.

Panofsky and Donovan's study draws its data from a single racist website, albeit the largest and longest-standing. Regardless, the utility of consumer genetic testing is now a major and significant part of White supremacy discourse. In all the categorization of different types of responses—good news, bad news, refuting bad news, condemning it, or brushing it aside—none of them resulted in epiphanies that they might change their scientifically illiterate views about race.

This at least demonstrates quite clearly Swift's maxim that you cannot reason someone out of a position they did not reason themselves into. In these cases, modern genetics is being misused as a crutch to support a political ideology, and to have that crutch removed by reality does little to topple the ideology. At least, I suppose, we can take solace from these incidents, as they show that the racism expressed by White supremacists is not supported by science. Amusing though they might be, most of the discussions around genetics and race in these forums concern those whose results come out with only northern European "White" DNA, and are bolstered by a misunderstanding of tests whose services are sometimes marketed and simplified to the point of being scientifically questionable.

To condemn commercial ancestry testing services because they are co-opted by racists is unfair. But it is the same warping of science that fuels both racists and typical hobbyist genealogists alike. Genealogy and genetics have a close but not perfect relationship. DNA can tell you some interesting things about family history and ancestry, but its powers are profoundly limited by fundamental biology, and the behavior of people, which is that we move and reproduce with remarkable breadth. Traditional genealogy has its own complementary limitations: Paper trails go cold for most families after only a few generations into the past. For most people, the shortcomings of these genealogical techniques are brick walls that cannot be hurdled. Genetic ancestry tests may be fun but, in my opinion, mostly offer nothing much more than a gaudy bauble.

You are not your genes, and you are not your ancestors. Most of your ancestry is lost, and can never be recovered. We can be clear on this with absolute certainty: You are descended from multitudes, from all around the world, from people you think you know, and from more you know nothing about. You will have no meaningful genetic link to many of them. These are the facts of biology.

PART THREE

Black Power

The last White man to win the hundred-meter final at the Olympics was Allan Wells in 1980. It was the Moscow games, and owing to the intensity of the Cold War, the US had boycotted, and their elite sprinters were absent. Including Wells, there were five White men in that starting lineup, as well as two Cubans and a Frenchman of African descent. The bronze medal was also taken by a White man, Petar Petrov, a Bulgarian whose personal best was 10.13 seconds. Though unknowable, it is likely that had US athletes been present, Wells, a Scot, would not have made the final eight, as his personal best was 10.11.

Not only was this the last time a White man won the Olympic hundred meters, it was the last time that White men *competed* in the final and the last race in which the winning time was above ten seconds. Since that pistol fired in Moscow in 1980, fifty-eight sprinters started the hundred-meter final. As I write these words, more than a year ahead of the COVID-19-delayed Olympics in Tokyo set for 2021, I can say that I'm confident that the winner will again be a dark-skinned man of recent African descent.

The men's hundred-meter final in the Olympics is the most prestigious race on Earth. Every four years, it is the formal measurement of the fastest a human can run over the shortest agreed distance, on the biggest stage available, and billions look on. The huge growth in the popularity of sport in the modern era, combined with global mass media, has meant that we can see people of every nation, every color and creed, compete in myriad competitions. The Olympics holds principles of international unity at its heart. Those iconic five interlocked rings on the Olympic flag represent the five continents—Europe, Asia, Africa, Australasia, and the Americas. The colors of the rings in the modern era are now nonspecific, though prior to 1951, Europe was explicitly linked to the blue ring, Australasia with the green, the Americas red, Asia yellow, and Africa black.

There are noble principles at the heart of modern sporting contests. The Olympic motto is Faster, Higher, Stronger, and it is a spectacle to showcase talent, hard work, healthy competition, and the struggle not for victory but simply to have taken part. As viewers, we see great entertainment in people at their physical zenith locked in the drama of intense conflict bound by strict rules.

However, these honorable values mask a lot of inequity. In sport, there is vast inequality of opportunity and thus of outcome. Not everyone has access to the same facilities and riches required to be successful in sport. Not all children have parents or caregivers wealthy enough to be able to sacrifice hour upon hour, day after day so that they can put in the training required to compete. Not all countries have the same cultural interests in specific sports. And as for the fundamental biology, sport, far from being a great leveler based solely on practiced skill and hard-won effort, is enormously skewed by innate physicality. This is obvious in the most basic way: Tall people have an advantage in basketball, and height is heavily and overwhelmingly determined by genes. Different body shapes suit different sports and even different positions in the same sport. An offensive lineman in football benefits from being a hulking beefcake, whereas a wide receiver needs to be lithe and fast like a sprinter.

These are traits that are significantly influenced by genetics, and so when we see the dominance of one group of people in a particular sport, we have to address the temptation to attribute their advantage to their ancestral origins. Here, I will anatomize two specific athletic domains to which this idea has been attached: sprinting and long-distance running. I will be predominantly referring to male competitions, as more data is known about men's sport and sports physiology. In reference to records, the men's times in elite races are faster than women's. However, there is no reason to think that any of what follows could not apply to women's sport as well.

The dominance of Black athletes in the modern era of sprinting has fueled a commonly held belief that people of African descent, and specifically West African heritage, are genetically predisposed to having physiologies that render them naturally at an advantage for sprinting. Although it is only in the last forty years that Black men have achieved total dominance in the hundred meters, underlying racist sentiments about the physicality of Black athletes are much older. In 1936, James Cleveland Owens (better known as Jesse on account of his Alabama pronunciation of his initials) carved out one of the greatest athletic achievements of all time by winning Olympic gold in the hundred meters, two hundred meters, four-by-one-hundred-meter relay, and the long jump. Better still, he did it in Berlin, which much vexed Adolf Hitler, who witnessed Aryan inferiority trail after Black dominance. A powerful photo exists of the aftermath: Jesse Owens saluting the American flag on the podium, surrounded by thousands of people extending their right hands in a Nazi gesture.

Our schadenfreude is undermined by the comments of Owens's own coach Dean Cromwell, who later said: "The Negro excels in the events he does because he is closer to the primitive than the White man. It was not so long ago that his ability to sprint and jump was a life-and-death matter to him in the jungle."

Attribution of sporting success via ancestry is a common trope, but has been applied unevenly. In the early twentieth century, Finnish people utterly dominated long-distance running, principally a superstar of early track and field called Paavo Nurmi, who won nine golds at three Olympics and set twenty-two world records. Jack Schumacher, a German writer asserting White superiority in the 1930s, used almost exactly the same argument as Dean Cromwell to justify the dominance of the Flying Finns, as they were known: that it is innate, inborn. In his interpretation, it is romanticized as Völkisch purity rooted in their natural terrain: "Running is certainly in the blood of every Finn. . . . Nurmi and his friends are like animals in the forest. . . . Their awe-inspiring times are a way of giving thanks to Mother Nature."

Paavo Nurmi racing at the 1920 Summer Olympics in Antwerp, Belgium

In contrast, attempts to explain the modern superiority of Black athletes invoke another cause for the selection of brute physiologies: slavery. Strength and power would be a desirable trait in enslaved men and women, so the argument goes. Individuals with those innate characteristics would've been successful in their bondage, and would be kept, traded, and rewarded. Therefore, they lived longer and had more children. Hence an unnatural selection would have increased the preponderance of these power genes. In January 1988, a famous football television commentator named Jimmy Snyder said:

> The Black is a better athlete to begin with, because he's been bred to be that way . . . they can jump higher and run faster because of their bigger thighs. And he's bred to be the better athlete because this goes back all the way to the Civil War, when, during the slave trading, the slave owner would breed his big Black to his big woman so that he could have a big Black kid. That's where it all started!

Snyder was fired from his twelve-year role on CBS the following day.

Michael Johnson, one of the great sprinters in the modern age—and for the record, my personal favorite track and field athlete of all time—said something in a similar vein in a television documentary in the run-up to the London Olympics in 2012. On uncovering his West

Michael Johnson sprinting off the starting line at the 1996 Summer Olympics in Atlanta, Georgia

African heritage via genetic tests, and learning about the brutality of transatlantic slavery, he commented:

> All my life I believed I became an athlete through my own determination, but it's impossible to think that being descended from slaves hasn't left an imprint through the generations. Difficult as it was to hear, slavery has benefited descendants like me—I believe there is a superior athletic gene in us.

This is an interesting argument, and is worth scrutinizing. There does appear to be some evidence for genetic differences in African Americans compared to Africans. These include the increased frequencies of genes that pose higher risks for hypertension, prostate and bladder cancers, and sclerosis, and lower frequency of alleles that cause sickle cell disease. There is no proposed explanation based on selection for an increase in the disease-associated genes, but the difference between African Americans and West Africans may simply be accounted for by admixture with Europeans since the introduction of slavery. A plausible mechanism for the lowering of sickle cell alleles might be that malaria is not endemic in large parts of the US where enslaved Blacks lived, though this is a short time period to account for this difference.

Or it might be chance. Those genetic differences may not necessarily be due to selection at all. They may merely reflect the fact that African Americans have a different migratory story from Africans, and those changing gene frequencies reflect different life histories. The idea that there has been evolution via artificial (as opposed to natural) selection specifically for physical prowess has a number of problems. Two or three centuries is not a very long time in evolutionary terms, and arguably not enough time for these genes to become fixed in a mixed population as a result of deliberate selection. Indeed, one 2014 study of the DNA of 29,141 living African Americans showed categorically no signs of selection across the whole genome for *any* trait, in the time since their ancestors were taken from their African homelands.

Breeding programs by slave owners did occur, but not uniformly or consistently. Furthermore, there were different types of slaves in America, what Malcolm X termed "field negroes" and "house Negroes" for whom physical strength would have not necessarily been a selective advantage. Furthermore, the economics of slavery were not a uniform industry to be served by one type of human chattel. Tobacco cultivation dominated much of agriculture in the South, but eventually gave way to cotton farming in many areas, which was far less labor intensive, and highly skilled. Powerful workers would not necessarily have been quintessentially important. I am unaware of any breeding programs specifically for speed.

Let us speculate, generously. Maybe selection during slavery *is* the biological difference between overrepresentation of African American athletic success compared to African. Let's leave aside the lack of support for that idea from generational time, and the absence of evidence for selection in the genome, as mentioned above. Let's pretend that the genes being selected relate to power and strength, and by extension, that translates into a sprinting advantage, even though slave-breeding programs were not for fast running. Why then are Eastern Europeans dominant in weightlifting, and absent in sprinting, when slavery selection for power would be perfectly attuned to this sport, much more so than running? Why do African Americans dominate in boxing, but not wrestling? Why is it that a game such as squash, which also requires explosive energy and power, is dominated by athletes from India, Pakistan, Egypt, and Great Britain, and has never featured a successful person of African descent? Why are there no African American sprint cyclists?

Tennis is a sport requiring strength and explosive energy, yet people of West African or any African descent are largely absent from this sport of privilege. With twenty-three Grand Slam titles (and an additional sixteen in doubles), the dominance of Serena Williams in modern tennis puts her as one of the greatest tennis players of all time, and indeed one of the greatest sportspeople of all time. Is Williams's success a result of her ancestry? Yes, in a narrow sense, in that her genetic makeup presumably bestows part of her advantage. But the question is this: Is her ancestry the defining characteristic of her success?

That a Black woman is a true great is partially a reflection of the lowering of prejudice and raising of opportunity in the modern era. By being one of the greatest tennis players of all time, just as Usain Bolt is the fastest runner ever recorded, they are already wonderfully freakish outliers and poor representatives of normal humans. Are they outliers genetically?

For sprinting, there is a notable and blindingly obvious fact that is forever ignored. African American,

Caribbean, and African Canadian athletes have dominated sprinting for forty years, all descended from the enslaved from West Africa. Only five White men have competed in the Olympic hundred-meter finals since the starting pistol was fired in the 1980 race, and none since those five from 1980 crossed the finish line, and the gold and bronze in that race are the only medals not won by Black hundred-meter sprinters. In that same time, the number of African men in the finals is also five. This includes two medals, both won by Frankie Fredericks from Namibia, a country that is not considered West African (rather it is southwest African); only one of the five Africans logged a time less than ten seconds. By this metric, African men are precisely as successful as White men. The transatlantic slave trade also imported millions of West African women and men to South America. The number of South Americans of any ancestry to have competed in the hundred-meter finals? Zero.

The point is this: Elite sprinters in the Olympics are not a dataset on which a statistician could draw any satisfactory conclusion. Yet it is precisely the data on which an extremely popular stereotype is based. The idea of Black athleticism in sprinting is drawn from a hugely skewed and fatally flawed sample, one that, owing to the relative absence of West African sprinters, doesn't even support its own hypothesis. If people of West African ancestry have a genetic advantage, why are there few West African sprinters, when slavery does not account for the difference?

We can of course go beyond mere speculation of evolutionary change and assess the molecular biology of physical abilities. The real genetics of sporting success are predictably complex. As with any human behavior, there are myriad factors in the physiology of physicality: the size of your heart; the efficiency with which you absorb oxygen (maximal oxygen uptake, also called VO_2 max); muscular recovery from exercise or injury; the lactate inflection point, which is when the levels of lactic acid shoot up owing to being produced faster than the body can break it down, resulting in muscle cramps or a stitch. These are all relatively well-understood phenomena that have a solid genetic basis. There are also physical traits such as flexibility and coordination that are less well understood from a genetic perspective. And finally, there is the psychological—determination, concentration, perseverance, risk taking—which, just like all behavioral traits, have a genetic basis, but are immensely complicated and poorly understood (see part 4).

This is a typically messy picture to unpick, so let us deal first with what we know the best. Power and stamina are at opposite ends of the spectrum of muscle performance. We know this intuitively: Elite endurance athletes and sprinters make a Venn diagram that does not overlap. We know it genetically, too. The contemporary approach to identifying the genes involved in athleticism is to take elite athletes and look for gene variants that are more common in them than in the rest of the population. With those differences, we can infer that those genes boost performance, without knowing what the genes actually do. This is a pretty standard technique in genetics, and is fruitful, too. More than 150 individual points of genetic difference have been identified in eighty-three genes in elite athletes in hundreds of studies, of which approximately three-fifths appear to relate to endurance and the rest to strength or power.

It is worth noting that in some of the elite athletes in power-dominated sports who were tested (rugby, kayaking, wrestling), gene variants were identified that fell below the threshold of being significant, meaning that they were probably not more commonly found in the sportspeople than in a broader public. While this does not undermine the fact of genetic advantage in sporting success, it highlights the importance of nongenetic factors.

So the question becomes this: Of the multitude of genetic variants identified so far that associate with elite sportspeople, do they segregate with specific populations, ethnicities, or races?

The answer is yes. And no. And maybe. We don't know the effect of most of those 150 variants, and we have some information about how they are distributed around the world. Here, I will focus on two in particular, both of which are heavily studied, apparently important, and also the subjects of a lot of hokey science.

Muscles are made up of long fibers built from multiple tubular cells. When you flex your biceps, all those cells spark into action and contract in unison to tighten along the length of the muscle, and draw the forearm in. Skeletal muscle cells come in two types: slow- and fast-twitch. Slow-twitch cells are more efficient at processing oxygen to generate the energy to contract than fast-twitch, which generate energy more quickly. Hence, fast-twitch cells are better for producing explosive energy over shorter timescales. People who are good at sports that require explosive energy tend to have a higher proportion of fast-twitch muscle cells.

The genetics that underlies this distinction is not well understood, though certainly involves a gene called alpha-actinin-3 (*ACTN3*), which, like all genes, comes in a number of different versions (or alleles), each subtly different. Two alleles correlate with much of the difference between fast and slow, and the difference is referred to as R577X.* Many studies have shown that elite athletes in

* Recall that a gene encodes a string of amino acids that make up a protein, and you have two copies of almost all genes. The R577X variant means that one genetic change at position 577 in the protein turns the amino acid located there from arginine (R) to a STOP (X) codon, resulting in a shorter ACTN3 protein in the muscle fiber.

power and strength sports are more likely to have one or two copies of the R type, rather than two copies of the X type, which results in fewer fast-twitch cells.

As we are invariably discovering in modern genetics, genes have many effects, and rarely can single attributes be ascribed to them. *ACTN3* is frequently described as the "speed gene," in both the popular press and in academic papers. Studies also show that the R allele is involved in response to resistance training, reduction in muscle damage after intensive exercise, and a decrease in risk of injury, but may be associated with reduced flexibility. It is worth noting that despite a strong interest in this gene from sports scientists and geneticists, its relationship with performance is not well understood. We have some demographic data though, and we know that the distribution of people with the XX genotype is globally uneven: A quarter of Asians are XX, a fifth of White Americans, one in ten Ethiopians, one in twenty-five African Americans, and only one in a hundred Kenyans.

So, the presence of the R allele (either one or two copies) is definitely higher in African Americans compared to White Americans, 96 percent compared to 80 percent. The numbers are almost the same for Jamaican people. That doesn't come anywhere near the observed discrepancy between African American or Jamaican Olympic sprinters and White competitors. If it were just down to that one gene, you might expect to see maybe six elite sprinters being Black for every five White runners.

Take another sport where explosive energy and speed are an asset: basketball. In the National Basketball Association, the ratio of Black to White players has been consistently around three to one since the 1990s, again Black people being significantly overrepresented if the R allele is your sole criterion. This is an ultra-simplistic argument, as obviously many other factors that are genetically influenced are important in basketball, notably height. In other sports, desirable body form is more variable. In the National Football League, the proportion of Black players is around 70 percent, but like rugby, that is a game where there are highly specialized positions with different skills and physical attributes. Offensive linemen tend to be heavy and strong, running backs tend to have the physique of sprinters, and most are Black. Linemen though are a fairly even split of Black and White Americans. But in the center position within the linemen, Whites outnumber Blacks four to one. Why? We don't know, but it does not appear to have anything to do with genetics. In Major League Baseball—a sport that requires sprinting and powerful throwing and hitting—African Americans make up less than 10 percent of players.

None of the numbers makes a great deal of sense if biological race is your guiding principle, and patterns in relation to ethnicity are terribly inconsistent both between sports and within them. And while there is uneven distribution of the R allele in different populations, this does not match the makeup of elite athletes in different sports.

Kenyans and Ethiopians account for around two fifths of the honors in the Olympics, World Athletics Championships, and World Athletics Cross Country Championships at middle- and long-distance running. Since 2010, *every* winner of the London Marathon, both women and men, has been either Kenyan or Ethiopian. The dominance of these two countries in endurance running on the highest stage is close to absolute. Why could this be?

Just as there is an assumption that West African ancestry is essential for dominance in sprinting, there is a persistent belief among many that East African ancestry is essential for elite success in endurance racing. Because of the geographical specificity of these elites, the accompanying suggestion is that there is an evolutionary basis to East African running success. Unlike the false assumption that selection via slavery drove the necessary genetic changes for strength and power, for endurance racing various ideas have been put forward, including that pastoralist ancestors in the East African highlands evolved to chase down their herds.

Body shape is a factor in endurance running success. Light and lean bodies are better at dissipating heat, and these physiques abound in East Africa, quite possibly adaptations to the local hot climate (in contrast to Tibetan or Inuit body shapes that tend to be shorter and more rotund to retain heat in the cold). The genetics that underlies endurance physiology is similar to but different from the muddle of *ACTN3*. The gene that is most heavily studied in relation to endurance sport encodes a protein called angiotensin-converting enzyme, or ACE. It sits on the surface of cells in the lungs, kidneys, testes, and other tissues, and is involved in the body's system for regulating blood pressure, by helping to control volumes of water flowing in and out of cells. The *ACE* gene comes in two major alleles, one with a chunk of DNA missing (called D; the longer version is called I). Both versions work fine, but the D form causes blood pressure to rise more quickly. People with the I form have higher oxygen uptake and higher maximal heart rate. In a meta-analysis of 366 studies (that is, one that aggregates multiple studies to increase the statistical power), the presence of two *ACE* I alleles was significantly higher in endurance athletes compared to ID or DD.

Predictably, the *ACE* II variant occurs at high levels in elite athletes from Kenya and Ethiopia. This is unsurprising, because when the *ACE* gene has been assessed in studies comparing Ethiopian and Kenyan elite runners with nonathletes from the same countries, no difference was found, meaning that for East Africa, it is a national genetic characteristic, irrespective of athleticism.

But on closer inspection, this might not be such an informative question to ask, because the populations from which elite endurance runners emerge are much more restricted. In fact, the specific demographics of success from these two countries are incredibly precise. For Ethiopians, the majority of international athletes come from the Arsi and Shewa districts. For Kenyans, it is people from the Kalenjin linguistic-ethnic group—surnames beginning with "Kip-" typify these people, such as the great runners Moses Kiptanui, Helah Kiprop, Wilson Kipsang Kiprotich, and Eliud Kipchoge, who in October 2019 became the first human to run the marathon in under two

Eliud Kipchoge, whose finish time was 1:59:40 at the Vienna City Marathon

hours. Even more specifically within the Kalenjin, the Nandi subtribe are disproportionately successful. Nandi and Arsi are mountain districts in the Rift Valley, more than 6,500 feet (2,000 meters) above sea level.

Physiology that works well at high elevation is advantageous in sport. There is less oxygen up high, so if you can cope with that, you have an edge when competing at sea level; there, oxygen levels are higher and you will be able to pump energy in your muscles with greater efficiency. Living and training at altitude is therefore a good thing for athletic success—athletes become accustomed to exercise with lower oxygen, and then get a boost at sea level. For the question of ancestry, a long-standing population who live at high elevation may well be necessary, but it is not sufficient to account for athletic success. If it were, we would expect to see great Mexican, Andean, and Tibetan runners. Large parts of South America, central Asia, and Mexico are similarly above a 6,500-foot (2,000 meter) elevation.

But they do not have a culture of running. And that is a key difference. In Kenya and Ethiopia, running is an industry. Successful coaches, bolstered by successful iconic runners, have set up intensive camps built on a culture of success. The Ethiopian mountain town of Bekoji, population sixteen thousand, has produced ten Olympic medals and fifteen world records. If this were a US equivalent phenomenon, it would be as though all

American Olympic gold medalists in athletics for the last fifteen years had come from the town of Sheridan, Wyoming—the 424th most populous micropolitan area in the US, out of 550.

In Kenya, the town of Iten is similar to Bekoji: intensive, expert, and highly specialized training from a large pool of motivated athletes desperate to be the next world record contender. Some have suggested that part of the genesis of this tradition can be traced to colonialism, with missionary and military influence promoting exercise. Maybe there is some basis to this, but famous runners such as Kipchoge Keino (gold 1,500 meters, 1968 Mexico) and Haile Gebrselassie (gold 10,000 meters, 1996 Atlanta) had transformative effects on running culture in their homelands.

As in all sports, the motivation to train hard and be part of the culture of running is also to enjoy the spoils of success. Winners earn good money and become celebrities. International talent scouts haunt the training camps to discover new superstars. When trying to account for the supremacy of Kenyan and Ethiopian runners, a 2012 study concluded that on top of the *ACE* II allele, body shape, metabolic efficiency, and intensive training (specifically to both live and train at altitude), there was also a strong distinctive "psychological motivation to succeed athletically for the purpose of economic and social advancement."

The genetics of East Africans is significant but is not unusual either nationally or internationally. A study of 1,366 people in London showed that the frequency of I and D alleles of *ACE* was the same in people of European and African descent, but the proportion of South Asians having two I alleles was significantly higher.

Just as *ACTN3* is not a speed gene, *ACE* is not an endurance gene. These simplistic reductions of biochemistry betray not just the complexities of their roles in the body but how much or little we know about those functions. "Necessary but not sufficient" is a phrase that geneticists like to use a lot. There is no reason to suppose that the variants of both *ACE* and *ACTN3* that form part of the foundations of elite athletic ability are unique to Africa or recent African descent. Are fast-twitch muscle cells more common in sprinters? Yes. Are they more common in West African people? Possibly. Are they more common in African Americans? Maybe a bit. Are they unique to African people? No. Does the RR allele of *ACTN3* or the II allele of *ACE* make you run faster? No: In elite athletes, they appear to be necessary but not sufficient for athletic success. The difference in regionally mediated success is culture. The utter dominance of Finnish long-distance runners in the first half of the twentieth century ended because the culture of running dissolved. The current dominance of Kenyans and Ethiopians in long-distance running, and descendants of the enslaved in the Americas in sprinting, is because they have cultures and icons of total supremacy.

The study of these two genes has been extraordinary, not least because sport is big business, and understanding sporting success is interesting. As with commercially available genetic ancestry testing kits, plenty of companies have sprung up offering direct-to-consumer (DTC) tests for both *ACE* and *ACTN3* genes, supposedly in order to steer young athletes' basic biology toward specific sports. But the murkiness of our current knowledge of genetics is such that the International Federation of Sports Medicine identified thirty-nine companies offering these tests and issued a statement in 2015 decrying their use:

> The general consensus among sport and exercise genetics researchers is that genetic tests have no role to play in talent identification or the individualized prescription of training to maximize performance. . . . In the current state of knowledge, no child or young athlete should be exposed to DTC genetic testing to define or alter training or for talent identification aimed at selecting gifted children or adolescents.

This is the current state of affairs in relation to all humans, regardless of ancestry. Maybe there are probabilistic predictions one could make about ethnicity and sporting success based on genetics, but they would be weak, at best. As ever, human genetics is as complex as human history, because human genetics is part of human history.

There is a real danger here of fetishizing two genes out of twenty thousand, in a way that steers us back toward an essentialist view of racialized sport. Many studies have shown that the versions of *ACTN3* and *ACE* we see in African American and African athletes are far from unique, and a 2014 study concluded that they "do not seem to fully explain the success of these athletes. It seems unlikely that Africa is producing unique genotypes that cannot be found in other parts of the world." Even with all that genetic advantage in place, we have to resolve again that having the right genes is necessary but not nearly sufficient to account for the dominance of any group of athletes in any sport.

Sport is a complex social and biological phenomenon, which, like all human activity, includes significant input from nature and nurture, meaning genes and everything else. It is effectively casual racism to suggest that biological ethnicity is more important than other factors, not least because it is virtually impossible to pick apart all the elements of a lived life to assess the ingredients of a successful recipe. In science we look to Occam's razor (or scientific parsimony) to understand phenomena, the concept that the best hypothesis is the one that requires the fewest assumptions. Although it appears to be a simpler answer, the claim that traditional racial categories are the cause of sporting success actually requires far more explanation than the following: There is some genetic advantage in sporting achievement, some of which will manifest as physical traits that tilt the balance toward success. Precisely accounting for that genetic advantage is impossible, and none of it segregates with the colloquial races.

The stereotypes and myths might have a semblance of grounding in personal observation, and tend to revert to some version of essentialism—that there is a singular signature that demarcates difference. But from a sociological point of view, these types of folk analyses tend to invoke deep prejudices that we may well be unaware of, and ongoing structural racism. The sociologists Matthew Hughey and Devon Goss analyzed hundreds of sports reports in the press over eleven years in the twenty-first century. They found that a biological basis of race was a common theme in describing sporting success. In comparing references to success by Black and White athletes, they found that innate physical ability was typical in descriptions of Black athletes and intellectual prowess or industriousness was the most frequently referenced criterion for success in Whites. The fixation on individual genes in the analysis of athletic success says inherent biology, not effort, is the mediator of success. Our cultural biases clearly say "Black brawn and White brains."

The association between physicality and race extends beyond sport, and into sex. There is the widely held belief that men of recent African descent have larger penises than men of other populations, and that men of East Asian descent have the smallest. The most recent (2014) and largest meta-analysis found no indication in more than fifteen thousand men that penis length or girth correlates with any particular population, racial category, or ethnicity. Part of the persistent racism directed particularly to people of African ancestry is focused on their bodies—physicality, power, sexuality—this being the theme of the horror masterpiece *Get Out* (2017). Rarely is success attributed to intellect or hard work. Again, this recapitulates the sentiments of the Enlightenment thinkers who founded the pseudoscience of race; even the positive attributes delineated by race are reflections of a lesser evolution.

There's also the pervasive belief that Black people excel at specific track and field sports because they require little specialist equipment—the patronizing myth of African long-distance runners being trained by running to their schools or because they ran without shoes and thus learned good technique. To become a top runner or soccer player, this implies, you just have to run, or kick a ball around. This is another form of soft racism, and is similarly not rooted in fact. When successful Kenyan runners were actually asked if the story of running to school was true, most said no—they walked or took the bus like other kids. The elite athletes we see in the Olympics or the FIFA World Cup have been carefully selected over many years of filtering through the most advanced training programs on Earth to get them to that level of excellence. To suggest otherwise is just a recapitulation of the idea of a people being somehow "closer to nature," which is meaningless blather.

What is very striking about all these radicalized attempts at assessing sporting advantage is how inconsistent they are. The genetic and therefore physiological advantage that supposedly equates to sprinting success appears to have no impact on short-distance swimming—the Olympic swimming equivalent of the hundred-meter sprint is the fifty-meter freestyle. Since 1980, when the last White men ran in the Olympic hundred-meter final, there has been only one Black man to compete in the fifty-meter freestyle final—Cullen Jones, who won the bronze medal for the US in 2016. Some have attempted to justify the absence of Black people in swimming by asserting that they have denser bones, and therefore are not as buoyant. For this there is no evidence, and yes, it is as ludicrous as it sounds, but it is such a persistent idea that it has been said to me by Black friends. A year before Jimmy Snyder offered his biological essentialist explanation of sprinters' success in 1988, another sports presenter, the former baseball player Al Campanis, asserted on the popular US program *Nightline* that Black people's lack of representation in swimming was "because they don't have the buoyancy." Like Snyder, Campanis was sacked the next day.

It seems absurd to say it, but the pivotal element in being able to swim is learning to swim, rather than contesting some imaginary biological sinking factor. According to USA Swimming, the official national swimming body, 64 percent of Black children in America cannot swim. A 2017 survey identified the most significant correlates with this statistic, including: a low number of parents and friends who swim, economic disadvantage (swim teams are mostly extracurricular and therefore have added financial costs), a lack of access to pools, and the absence of African American swimming role models.

It is impossible to separate these trends from the racist history of America. Even after segregation officially ended in 1964, pools were more likely to be built in predominantly White areas, and access to pools for African Americans continued to be restricted. It is the social and cultural milieu that is most significant in the almost complete lack of representation of African Americans in a sport that in terms of sheer biological physicality is no different from a sport in which they reign supreme.

The real-world consequence of this structural and cultural racism is that the death rate from drowning in

African American children aged five to fourteen is three times higher than for White children. Racism is literally lethal.

Sport is one of the ways that we measure physical and psychological excellence in people from all over the world. Because of nationhood, elite athletes represent countries, and therefore they represent us. We can play out all manner of battles, stereotypes, and prejudices in that arena. The science that underlies success at the top of any game is inscrutably complex, partly because it's like trying to unbake a cake, but also because though elite sportspeople represent us, they are not really like most people. They do things most people cannot. The genetics of our physical traits are immensely complex, too. They reflect individual differences, population differences, regional adaptation, and the weirdness of human history.

Nevertheless, we impose all our hopes, dreams, and prejudices onto our elite athletes, and these include deep cultural biases, many of which we may be only partially aware. It seems possible to pick almost any argument, racist or otherwise, and use sport to defend it. With such limited data, these positions are quite untenable. As well as entertainment, sport is a celebration of the extremes of human capabilities. To reduce it to mere unearned biology is racism, whether conscious or not. In return for their pursuit of greatness, we owe elite athletes more deserving praise than auspicious ancestry.

PART FOUR

White Matter

Here are some facts. There's about three pounds of meaty tissue inside your skull, which means that you have a large brain. Ours are not the biggest among animals, as brains scale with body size, and the human brain is a little nut compared with that of a blue whale. Our brains are large for our body size, but that ratio is much greater in ants and shrews. Ours are densely packed with specialized cells in our cortex, where most of our higher functions are seated, but crows have similarly dense neurons. Yet humans are special, and all of our consciousness, thoughts, imagination, and experience of the universe happen in that lump of gelatinous matter between our ears. But the basic biology of our brains is not fundamentally different from any other animal. Brains are part of our bodies, and our bodies evolved under the auspices of natural selection. We know very well that some of the physical form of human beings has adapted to suit the different environments that our ancestors spent time in: pigmentation, diets, exposure to diseases, elevation from sea level—these are all things that have crafted our bodies so that we would survive.

Given that brains are part of our bodies, could it not also be true that the very real different cognitive abilities that different humans display are also a result of molding by living in specific areas, with specific ancestries?

When it comes to some of the metrics of cognitive differences between the so-called races, the figures are stark: The number of science Nobel Prizes won by Jewish people currently is 144. The number won by Black people is zero. As in sport, performance at the extreme ends of achievement is not necessarily reflective of the population whence the winners came. When it comes to measuring cognitive abilities, we typically look to population averages. There, the numbers are no less edifying: According to some studies, Black populations around the world do less well in IQ tests by a margin—some estimates put the gap at between ten and fifteen points on average. The inheritance of intelligence is probably the most controversial topic in the whole of science, and when it is combined with the study of population differences, evolution, and race, there we have the prospect of a perfect storm. If you are using science to justify a racist opinion, observations of performance differences of these groups in cognitive tasks are the end of a conversation. For someone who is interested in science as a mechanism for pursuing truth, they are the beginning.

It is often said that this is a taboo subject, and that honest discussions around race, intelligence, and genetics are the preserve of brave crusaders who refuse to kowtow to intellectual censorship born of denying reality. A cliché that is bandied around asserts that scientists "sacrifice truth at the altar of political correctness." I don't recognize this picture, and it seems often to be conjured by people casting themselves in this aggrandizing heretical light—truth seekers versus those who pervert scientific purity. The mainstream press and online niches foment this polarization, frequently conflating ideas via glib phrases that serve to nurture this conflict—"virtue signaling," the term "snowflakes," and meaningless slogans such as "facts don't care about feelings"—all designed to evoke a sense that there is a culture war between a side that seeks only to reveal the hidden truth, and one that wishes to suppress it. Yet a superficial search for articles on race and intelligence will unleash a torrent. Far from being a taboo, one that supposedly violates principles of free speech, the topic unleashes a flood of discussions of race and intelligence, and this has been the case throughout much of the twentieth century. The magnitude of this current deluge doesn't speak of supposedly forbidden knowledge. Instead, we see a popular and sometimes academic discourse that is clouded by complexity, confounding factors, and—in that elegant Darwinian phrase—ignorance begetting confidence. As in the early Enlightenment days of scientific racism, a serious, complicated, and ongoing area of important research is being marshaled and co-opted into a political war zone.

This field is beset not just by ideological battles but by some mountainous scientific terrain—and we are currently only in the foothills. The brain is quite possibly the most complex object in the known universe, and the genome is the richest dataset that we have yet discovered. Simple answers therefore were never going to be forthcoming. There are several problems inherent to this topic. The first is that genetics is fiddly and hard, and we are only just beginning to figure it out. The second is that measuring intelligence is complicated and hard, and while we have plenty of metrics, there is plenty of scientific controversy within the reams of data. There is also the fact that race as colloquially described is not reflected in our genomes accurately, as discussed throughout this book. So linking these three concepts is far from neat: race and genetics, race and intelligence, and intelligence and genetics. They do not make easy bedfellows. Hundreds of books and thousands of papers have been written on these subjects, for more than a century.

Controversy is stoked by questions of race and intelligence, often prompted by racist comments by public figures. James Watson, co-discoverer of the double helix structure of DNA and champion of the Human Genome Project, repeatedly made racist comments for many years, both in public and in private. In an interview in 2007 he said he was "inherently gloomy about the prospect of Africa" on the grounds that "all our social policies are based on the fact that their intelligence is the same as ours, whereas all the testing says, not really." On the question of the equality of races, he said that "people who have to deal with Black employees find this not true." On one of the three occasions I met him, he told me that I was going to be OK in genetics as "Indians are hardworking though unimaginative." I had been working in science for nineteen years by this point.

In a 2018 documentary, Watson, by then old and infirm, indicated that he hadn't changed his mind, despite publicly apologizing in 2007 for those very same comments.

It is a shame that a life punctuated by truly great scientific achievement should end under the shadow of ignorant self-imposed ostracism. In 2019, Watson's lab at Cold Spring Harbor removed all his remaining titles and his portrait, as did other labs around the world. The platform he had earned though epoch-defining research had been eroded by his repeated expression of scientifically ignorant and straightforwardly racist views. Geneticists had finally had enough. But for a few people fixated on questions of race and intelligence, their ire was reignited, and Watson became a champion of the faux-persecuted, a man excommunicated from the very field he helped establish, for merely telling the truth. Except it wasn't the truth at all. Instead it was the repeated expression of fairly unoriginal racism that anyone who had met him was all too familiar with: Black people are lazy, Indians are industrious but unoriginal, Jews are intellectually superior, all views expressed readily in the nineteenth and eighteenth centuries, when the foundations of scientific racism were being cemented.

My view was and is that we should be capable of recognizing and celebrating great scientific achievements while simultaneously condemning bigotry, even when they occur in the same individual. Like Francis Galton, James Watson was a brilliant scientist, and a racist. The political fallout of his comments is for others to discuss, but the question about this particular statement of his from 2018 remains, and it is one that is often expressed: "There's a difference on the average between Blacks and Whites on IQ tests. I would say the difference is, it's genetic."

Was he right? In order to scrutinize the stark statistics of cognitive abilities and race, we need to try to understand how intelligence has been scientifically assessed over the last century, and what it actually *means*. And we have to examine the current state of play in understanding the relationship between intelligence and genetics.

These are treacherous waters. Intelligence is not an easy thing to define. Cognitive abilities cover a range of behaviors, but generally reflect adeptness at reason, problem solving, abstract thought, learning capability, understanding ideas, and so on. We are talking about human intelligence here, and so it is beyond the broader sense of intelligence roughly meaning *doing the right thing at the right time*, which might be applied to other animals. Bees and ants perform all manner of essential problem-solving adapted behaviors, from navigational dancing to tending to their dead, to farming nutritious fungi off carefully cultivated leaves. Bees are objectively better than us at making honey. Yet they are terrible at IQ tests.

Cognitive abilities, like pretty much every human trait, are not evenly distributed among people: However it is assessed, some people are more intelligent than others. The most frequently cited and best-known assessment is the intelligence quotient—IQ. This is a test and metric that has been around for more than a century, and though the assessments are not the same now as the original 1912 incarnations, there are several versions, and they are standardized to include tests of reasoning, knowledge, mental processing speed, and spatial awareness. In a typical test, you will encounter sets of shapes in grids of nine, each row changing a part of the shape in a sequence, and then the ninth slot is missing for you to fill in from a multiple choice. There are other reasoning tests, where you might be asked to resolve a logic puzzle:

> Alice is sixteen, and is four times older than Ben; how old will Alice be when Ben is half her age?*

And you will do spatial analysis tests, where you have to imagine rotating a 3D object, and pick from a multiple choice which is the correct outcome.**

I am the son of a psychologist, and I have done these tests more times than I care to count. Over the years, I have assuredly gained the hard-won knowledge that they are really boring. Admittedly, that's not a very sophisticated analysis, but there are many serious detractors of IQ, and those criticisms come in many forms, with varying degrees of sturdiness. Common arguments against IQ include the notion that the tests are culturally biased, or that they lack appreciation of practical intelligence or creativity. Another argument dispels IQ as merely a score that measures how good someone is

---

* See footnote on the next page.

** Beware of online IQ tests, which sometimes are free to take but require a fee or registration to get your score. These are often not very robust or scientifically valid.

at IQ tests. That of course is literally true in a narrow sense, but is also not a very clever thing to say. The hundred-meter sprint only tells you how good you are at running that distance as fast as you can. The driving test only assesses whether you are competent enough at driving to be legally allowed to. It doesn't assess your bicycle proficiency, or if you have the potential to be a Formula 1 champion. It is easier to measure something than to understand what it is you are measuring. But this does not invalidate the measurement itself, if performed honestly.

These criticisms are all true to a certain degree, but they aren't secret revelations: Psychologists are acutely aware of these limitations, and modern tests are designed accordingly, though not perfectly. IQ tests are culturally biased, but that doesn't mean the data generated is invalid.

Another frequently voiced criticism is that a single metric is a poor way of assessing an immensely complex and multifactored set of behaviors. That is also true for the hundred meters. The outcome is a single figure, which doesn't specify the amount of training, the genes you were born with, how long you've been an athlete, and a host of other things to do with physical and psychological abilities. But that single metric will correlate very well with those and many other factors: We could make all sorts of predictions based on your speed. A time of under

---

ANSWER FROM PREVIOUS PAGE: Twenty-four

ten seconds will correlate very closely with being a professional athlete who has been training for a long time, with a predisposition to possessing genes associated with explosive energy, and a low heart rate. It currently predicts being descended from the enslaved from Africa, as the majority of the 140 or so people who have run that time have been African American (as discussed in part 3, though of course this may change in time). It predicts that you have two legs, two arms, and don't smoke, because no Olympic hundred-meter finalist has smoked or had fewer than four limbs, at least at the time of the race.

IQ, regardless of precisely what it is measuring, makes a much better predictor of many more things than a sprinting time does, and that is primarily because IQ has been tested and scrutinized for a century in thousands of studies. That alone makes it a useful metric. As is often the case in science, IQ has great value when applied to populations, and less when applied to individuals. Stephen Hawking, not known for being a sodden-witted dunderhead, was asked in 2004 what his IQ was, to which he replied, "People who boast about their IQ are losers." President Trump on the other hand frequently talks of how high his IQ is, including declaring that it is above that of his two predecessors in that august seat of power. Membership of supposedly prestigious organizations such as Mensa use IQ as an entry criterion, but frankly, I can't imagine a less interesting group of people to hang out with. Historically, IQ has been used for far more pernicious ends than joining a self-congratulatory club, and this goes some way to explaining popular hostility toward this valid scientific tool. IQ testing in the US in the first half of the twentieth century was applied as part of the assessment for state eugenics policies, which resulted in the forced sterilization of more than sixty thousand people.

The way the test results are processed is that the average is set at one hundred points, and the range of IQ across a population falls into what is known as a normal distribution, aka a bell curve. This means that there is an equal number of people above and below one hundred, and that around two-thirds are within fifteen IQ points in either direction. About one in forty people is above 130 or below 70. IQ is not fixed during one's life though: Results tend to stabilize as you get older, but fluctuate wildly during adolescence. It can also be improved, marginally, with practice, notably when schools adopt different teaching strategies that are rewarded in standard IQ tests. This is hardly surprising, as IQ is a test of current skills, which are developed, rather than some innate immutable intellectual power.

Nor is IQ fixed through time in populations. There is a phenomenon known as the Flynn effect. The political scientist James Flynn observed that IQ was rising in test groups on average by around three points per decade from the 1930s onward. There are several factors that may account for this, including improved health, nutrition, standard of living, and education, but changes in genes have been ruled out. Because the effect is seen in many places around the globe, and has been observed in just a few years, substantive genetic changes cannot have occurred either within or between generations.

We see versions of the Flynn effect in other human endeavors, too. Athletes are generally fitter than they were in the past, by pretty much every measure. How would the 1920s Detroit Tigers with Ty Cobb or the Yankees with Babe Ruth at their relative primes fare against *any* major or even minor league outfit a century later? I'm fairly sure that England's World Cup–winning soccer team from 1966 would struggle against the current first team of my beloved Ipswich Town, at the time of writing languishing in the third tier of the English soccer leagues. Are sports teams now genetically better? Not substantively, but as sport has developed and become more serious and more lucrative, training programs, equipment, diet, fitness, and professionalism have all driven standards stratospherically.

The value of IQ for science is undeniable. It also correlates well but not perfectly with other measures of cognitive abilities that are often used in scientific studies, such as educational achievement (results in exams) and duration (how long you stay in education). People who score well in IQ tests tend on average to live longer, get better grades at school, be more successful at work, and have a higher income.

When it comes to looking at IQ scores around the world and between different populations, the picture is far from clear, but there are some undeniable differences. The most up-to-date meta-analyses suggest that countries in sub-Saharan Africa are likely to score in the eighties,* as compared to US IQ standards, though these results are not universally accepted. This, obviously, is significantly lower. Interpreting these results is not easy at all, and while it is not possible to fully exclude genetic factors,

---

*Earlier studies, notably led by the controversial psychologist Richard Lynn, suggested much lower IQ scores in Africa on average, across the continent, with population average results in some countries as low as the seventies. However, that conclusion has been justly criticized as having been deliberately drawn from carefully selected and unrepresentative data that significantly lowered the averages, with no real explanation as to why his studies were so unsystematic in which datasets were chosen. Lynn is on the political far right, has spoken at far-right conferences and events that have also hosted a former Ku Klux Klan grand wizard and other White nationalists, and he espouses secession for US states to preserve "White civilization." It should in principle be possible to consider Lynn's work independently of his racist and White supremacist views. But issues over questionable data and cherry-picking of results put heavy strain on his credibility as a scientist. "What is called for here is not genocide," Lynn reportedly said in 1994, "the killing off of the population of incompetent cultures. But we do need to think realistically in terms of the 'phasing out' of such peoples. . . . Evolutionary progress means the extinction of the less competent."

these seem unlikely owing to the immense genetic diversity that is now well established across that continent.

Rather, environmental factors are a much better fit for explaining the discrepancy. Developing countries have lower standards of living than developed countries, as well as less sophisticated education systems, health programs, and medical care. These sorts of things are not easy to quantify, and the data is sparse and unsatisfactorily averaged across multiple African countries, which are all different. But some IQ researchers have credibly suggested that the socioeconomic status of many sub-Saharan African countries is similar to that of European countries in the first half of the twentieth century. Indeed, the authors of the largest meta-analysis of IQ in this region point out that if the Flynn effect had not occurred in the Netherlands (for example), the Dutch national IQ would currently be as it was in the 1950s, that is, around eighty (compared to today). Similarly, one study, again not universally accepted, put the national average IQ in Ireland in the 1970s at around eighty-five, but it is now at one hundred, the same as in the UK, and a few points ahead of the US. Again, that change, if real, occurred within one generation, so genes cannot be the driving factor. Instead, profound socioeconomic changes happened in that short time; health and education improved and were heavily invested in, and rural agricultural lives rapidly gave way to richer and more complex urban and industrial culture with mass media. It could therefore be sensibly argued that a big part of the alleged discrepancy we see between some African and European countries can be attributed to the Flynn effect not having happened universally, and significantly not in some African countries. If the factors that have driven increasing average IQ in some populations include better nutrition, health care, and education, it is plausible that these have not improved significantly enough to fully close the gap. As IQ is such a strong predictor of matters related to quality of life, understanding the science that underlies these things is important.

We learn from our families; we inherit genes from our parents. The people who live near us tend to be more closely related than random strangers. Social policies operate at national levels, and combined, these factors narrow the geographical influence on how any human characteristic is transmitted through time.

Intelligence is highly heritable. That is a seemingly simple sentence to say, but in those four words is some of the hardest and most misunderstood science that we have yet attempted. Broadly it means that a significant proportion of the difference we see between people is accounted for by DNA. Height is a simpler trait to help understand what this tricksy concept means. On average, tall people have tall children. We know from twin studies (and other methods) that most of the difference in heights in a population is based in genes rather than the environment. If we were studying a group of people where the tallest person was seven feet tall and the shortest was five feet tall, the latest studies indicate twenty-two of those twenty-four inches of difference would be encoded in DNA, and the remainder would be variation caused by the environment—such as diet and nutrition. That doesn't mean we know what those genes are, or what they are doing, just that variation is encoded in DNA.

"Heritable" is a wretched piece of jargon, because it doesn't mean what it sounds like. Heritable does not mean how much of a trait is genetic and how much is environmental—nature and nurture. Here is another example: Let's say that all humans are born with ten fingers, five on each hand. At birth, there is no variance in finger numbers, which means that this trait is entirely determined by innate, genetic causes. But many adults have fewer than ten fingers, as they may have lost them in accidents. So the variance in finger number in adulthood is entirely determined not by genes but by the environment, and therefore the heritability of finger number in adults is very low, close to 0 percent.*

---

*In fact, as some babies are born with fewer or more than ten digits, some of these changes can be caused by environmental factors, such as in the case of the drug thalidomide, where babies were born with abnormal limbs and digits. What this means for the purpose of this analogy is that the heritability of finger number is very low, not undefined.

That's an extreme version to make the point, but almost all traits are heritable to some degree. Cognitive abilities by whatever measure are no different: Innate levels of intelligence are highly heritable. Tabula rasa—the idea that we are born with a blank slate on which our abilities and personalities are drawn—is not correct. And we've known this for decades. Estimates vary depending on the study, but the proportion of cognitive abilities that can be attributed to genetics rather than other things is somewhere between 40 percent and 60 percent. That means that roughly half the differences we see are due to differences in DNA. These are not particularly new findings, nor are they very controversial: The slate is not blank—it is partially written at conception with the DNA of our forbears.

Calculations about cognitive abilities have historically been done with techniques that include nature's most helpful experimental tool: twins. Identical twins have (almost) precisely identical DNA, so any differences between them in any behavior should be down to nurture not nature. Identical twins who were separated at birth are another version of this tool, as they will have been nurtured in different families. But there are limitations on and complications to these methods, which are certainly not crippling but worth bearing in mind: Twins separated at birth are likely to be raised in families in similar populations, in the same countries, and of course at the same time, meaning that the environmental differences may not be radical. Identical twins share twice as much DNA as siblings, but because siblings and twins share their environments, the heritability of traits in identical twins is not double that of nonidentical siblings. With these shortcomings, and others, twin studies are still a valid and important part of understanding the heritability of intelligence.

In the modern era we are now looking for the actual genetic differences that correlate with complex traits. We can scan through the genomes of hundreds of thousands of people and look for slight variations in the genetic code, and try to work out if they appear to cluster with particular behaviors. These are called genome-wide association studies, or GWAS (pronounced *gee-waz*). Since these were first invented and deployed in 2005, the GWAS has become a mainstay of genetics.

The great revelation of the Human Genome Project was that we don't have very many protein-coding genes; fewer than a water flea, a roundworm, or a banana. The count for human genes comes in at around twenty thousand (depending on how you define them). This meant that the traditional model held by many geneticists of "one gene for one trait" fell apart at the seams. Instead, for the last fifteen years or so we've been building a new model of how genetics works in us, and part of that revelation is that single genes frequently do many things in the body at different times. Genes work in networks and cascades and hierarchies. And so for traits that can be summarized in a simple metric—height, eye or skin color—what we find via GWAS is that a handful, dozens, or even hundreds of genes play a small but cumulative role.

IQ is a single number, but intelligence is not a single thing, and the genetic component to intelligence is most emphatically not a single gene. The most recent studies identify scores of genetic variants that correlate en masse with better results in cognitive tests. These differences are in genes that we all have, and the cumulative variance appears to be the thing that correlates with performances in tests. The number of genes involved is likely to go up as the resolution of the genome gets sharper and the sample sizes get bigger. I would be unsurprised if the number of genetic variants that associate with cognitive abilities hits the high hundreds if not thousands.

Our newfound knowledge says that human genes often do many things in many tissues; genes involved in metabolism might be active in cells in different tissues all around the body. Given the intense metabolic demands that the eighty billion cells in our brains exert when thinking, doing, and generally maintaining a living soul, it is not a surprise at all that thousands of genes are involved.

We don't know what most of those genes do, at least at the level of molecular precision. Nor do we know what or how slight variations in them might affect our brains or behavior. The *A* in GWAS stands for "association," which means that the studies are revealing statistical correlations and the mechanics of whatever is being investigated remain anonymous. A GWAS plants a flag in the map of the human genome that says that something interesting is happening here, but we don't know what it is. These unknowns do not invalidate the method or the results—a scalpel is a precision tool essential for anatomizing a heart, but it won't tell you what an electrocardiogram is designed to do. It is quite probable that many of the observed differences in DNA encode subtle and not very informative changes to a protein's activity.

Imagine two versions of the Bible, the King James (KJV) and the New International (NIV). They are the same book, with the same overarching messages and the same stories, but many of the spellings, words, and indeed sentences have been changed, edited, and omitted. Some of these changes are trivial: In Revelation 13:18, the infamous passage about how to spot the Antichrist, the KJV says: "Let him that hath understanding count the number of the beast: for it is the number of a man; and his number is Six hundred threescore and six," whereas the NIV says: "If anyone has insight, let him calculate the number of the beast, for it is man's number. His number is 666." Some changes are larger, and arguably of greater significance: The NIV of Matthew 20:16 says, "So the last will be first, and the first will be last," omitting the second clause present in the KJV: "for many be called, but few chosen." Biblical scholars may argue about the hundreds of differences in these two versions, whether they are changes to the translations that significantly alter the sense, merely simplify the text, or do nothing at all. But simply by reading these sections of text, it should be possible to identify which Bible you are looking at. The GWAS does something similar—it looks at the genetic text and calculates a probability that the changes are coincident with a human trait.

We often use books as analogies to understand genetics—letters, words, sentences, coherent meaning, these are all things that are conceptually shared in both biology and literature. But the truth is that at this level of complexity in human genetics, no analogy can convey the richness of the data, nor the number crunching required to take it apart forensically.

Understanding population genetics, however, is important for scientific arguments about race, so buckle up, and I'll be brief. For really complex human traits, where thousands of minuscule differences seem to have small but cumulative effects, aggregating them helps us pool the genetic influence. This is known as a polygenic risk score (PRS). It's a metric that allows us to estimate the total genomic underpinning of a trait: When the outcome of a GWAS is many genes, totaling up their effect is handy. It is a powerful instrument, and a valuable addition to the scientist's toolkit. Polygenic scores help us to understand the genetics of any human trait, including complex diseases, though not yet with enough detail to warrant clinical intervention.

GWAS and PRS are truly brilliant tools that have utterly transformed the field of human genetics. That does not mean that they are infallible as tools, nor that they are always the most appropriate tools. Polygenic scores potently affirm that intelligence is heritable within a population. But this is a tool that is not particularly adept at dissecting the differences *between* populations. So when we see different IQ scores in different populations, and we know that the heritability of intelligence is high (more than 50 percent), that doesn't mean necessarily that the different DNA variants account for the differences between the populations. It would be perfectly possible for two populations with different sets of genetic differences to get the same IQ scores. Take height again, as it is easier to measure and better understood: The many differences in DNA that go some way toward accounting for the differences we see in heights in a population are not genes *for* height, only ones *associated* with height. We don't know what those genetic signatures do, whether they are influencing height or just are physically linked to genes that are important for height, and are going along for the ride owing to the way DNA is chopped up when sperm or eggs are made. We also don't know if the genetic differences are dependent on the local environment to drive the phenotype and make the average taller. We might expect to see different genetic associations in Europe compared to Japan when looking at height, but without knowing what those genes do, we can't know whether they have drifted into being, are significant, or are really meaningful in different environments, with different food or nutrition. These are not questions that this type of genetic study can answer well. GWAS are important and powerful for finding genes of relevance within a population—but not between different populations.

This, I will not dispute, is difficult technical science and statistical analysis. But it is important in the ongoing discussions about intelligence and race. We get better at forensically unpicking groups of people, and the tools become easier to deploy. That doesn't mean the tools are the right ones for the job. Scientists studying human variation, and journalists and readers need to be wary of drawing wrong conclusions from right results. Since the birth of the GWAS, results have frequently been vastly misreported: weak correlations claimed as causes, typically in the format "scientists discover the gene for X." With the development of the polygenic score on top of the GWAS, we run the risk of further overinterpretation, or outright error. When trying to account for disparities in intelligence, we need to bear in mind not only the limitations of the tools at hand but also the reasons we attribute to our observations. In this specific case of intellectual performance, the question is: Has biological evolution via selection—natural, artificial, or both—driven the difference we can see between populations?

Let us set aside the crippling difficulties in describing populations as distinct, discrete, or as races, as discussed at length earlier in this book. "Black" is not a taxonomic term that usefully describes the genomic, phenotypic, or geographic variation seen in Black people, though we can predict with an increasing degree of certainty where a proportion of a person's ancestry came from based on their DNA. Jewishness is a different type of ethnic and cultural grouping, and has an unusual history, owing to millennia of persecution, multiple diasporas, and forced migrations around Europe and beyond. People who identify culturally as Ashkenazi Jews carry a genetic signature that broadly, though not exclusively, suggests Jewish ancestry.

Mark Twain wrote an essay in *Harper's* magazine in 1898, entitled "Concerning the Jews," in which he concluded:

> His contributions to the world's list of great names in literature, science, art, music, finance, medicine, and abstruse learning are also away out of proportion to the weakness of his numbers. He has made a marvelous fight in this world, in all the ages; and has done it with his hands tied behind him.

Around a third of Jews are Ashkenazi, which is the group mostly likely to be associated with success in those intellectual pursuits, and makes up the largest proportion of American Jews. There are around eleven million Ashkenazi Jews alive today, but their history is far from clear. The Ashkenazi emerged in central Europe in the Middle Ages, though more specific dates and locations are fuzzy. Migrations from the Middle East and into central Europe seem to play a significant role in the development of Ashkenazi as a distinct cultural group within Judaism, especially into southern Germany, Italy, and France; in some of those places during medieval times, there was compulsory wearing of the yellow badge to identify Jews. Expulsions from those countries and Britain also contributed to the pushing of Ashkenazi Jews east, into Poland and Prussia. These centers of Jewish populations were relatively stable and would form the basis for the majority of the six million Jews systematically murdered during the Holocaust. Following that genocide, Ashkenazim migrated to many countries including the US and

Canada, as well as Israel, where they make up around half of the Jewish population.

Part of this unusual history includes practices and restrictions placed on Jews by those in power that focused large parts of their professional culture on business and commerce. That, combined with a presumed relatively high degree of marriage within the same social group, forms the basis of attempts to explain the observed intellectual success of Ashkenazi Jews. The general argument suggests that artificial selection for genes concordant with cognitive ability has been enriched as a result of the unusual history of Jews, and that this genetic selection accounts for their relative success. Consequently, it is argued, there is a genetic predisposition for great success not only at the extremes of intellectual ability—disproportionately high numbers of Nobel laureates, chess masters, leading violinists, mathematicians—but also more generally. According to some studies, Jews score significantly higher in IQ tests, which is a population average rather than due to extraordinary outliers.

Though anti-Semitism is thousands of years old, and Jewish associations with intellectual pursuits are centuries old, much of the current discourse on supposedly innate Jewish cognitive abilities stems from a single study in 2006. In a paper that has had significant influence—and initiated much scrutiny—Gregory Cochran, Jason

Hardy, and Henry Harpending suggested that the history of Ashkenazi Jews in Europe had the effect of enriching genes associated with intellect.*

They propose a number of factors unique to Jewish (and specifically Ashkenazi) history that created these favorable conditions for intellectual selection. These include social behaviors, such as endogamy, meaning that they mostly interbred, which created a gene pool favorable to natural selection. And "they had jobs in which increased IQ strongly favored economic success, in contrast with other populations, who were mostly peasant farmers." "Winnowing through persecution" is another suggestion in the paper—somehow, acts of oppression

---

*Henry Harpending, it may be worth noting, was a well-loved anthropologist whose work was influential and important. But toward the end of his life he garnered a reputation for being a controversial figure, and not just because of this study. He sometimes associated with far-right organizations and said things that are arguably racist. At a conference in 2009 entitled Preserving Western Culture, Harpending said: "When you view your parents or grandparents, and you know that they're retired, they could relax. But afterwards they can't just sit on the couch and relax, they've got to go and get a shop and work on a cradle for their grandchildren . . . I've never seen anything like that in an African. I've never seen anyone with a hobby in Africa. They're different." Published and peer-reviewed scientific research should in principle be regarded separately from personal opinions—the double helix structure of DNA remains true regardless of James Watson's ceaseless effluence of racist views. Nevertheless, choices about what one researches are undeniably influenced by one's political opinions, and as with Richard Lynn's scientifically and politically questionable work, Harpending's expression of racist views did nothing to showcase neutrality in his research topics.

and tyranny resulted in survival of the smartest. The authors, however, are clear in the paper that they can't explain how that would work, as no such effect is seen in other persecuted people. I find it most strange that such guesswork is included in a scientific study. In response to the assertion that professions involving commerce require high levels of intellect, I am unaware of strong evidence that success in business correlates with significantly above-average intelligence. Cochran et al. describe moneylending and other forms of commerce that are presumed to be the preserve of Jews as "cognitively demanding jobs" and state that "the Ashkenazi niche was so specifically demanding of accounting and management skills." Presenting this as evidence also sounds pretty sketchy to me. Medieval moneylending is not exactly rocket science, and it's definitely not medieval brain surgery.

They also cite specific biological factors and "physiological effects that could increase intelligence." In the ancient days of 2006, we knew less than we do today about neuroscience, and how biochemistry in cells relates to thought and action—but not that much less. Neuroscience is a vibrant field, but the truth is that we still really have very little idea of how neuronal growth and connectivity relates to cognition. If I have succeeded in convincing you that genetics is bewilderingly complicated, apply that to the development of the physical brain and the esoteric nature of thought, and you face one of the great frontiers in science. The suggestion is that some disease genes have specific effects on the growth of neurons, in a way that might enhance IQ. But this reflects a profoundly simplistic view of neurological development.

This neurobiochemical speculation was only one of the defenses. The major part was the high rates of certain diseases in Ashkenazi people. These include conditions such as Tay-Sachs disease, Gaucher's disease, and Niemann-Pick disease.*

The model they are copying is that of sickle cell anemia, a terrible lifespan-reducing disease that is recessive, meaning that someone must have inherited two copies—one from each parent—of the mutated gene to have the condition. When a person inherits one copy of the mutated gene, it is referred to as sickle cell trait, which is not nearly as serious as sickle cell anemia, though still has some associated symptoms. The disorder is often thought of as being specific to Black people, and therefore an example of how biology recapitulates race. But this is not

---

*Tay-Sachs disease is a severe condition that results in nerve cells dying and babies losing physical abilities such as crawling or rolling over, as well as hearing loss, and frequently death in childhood. Gaucher's disease is a nonlethal but complex set of disorders that includes skeletal problems, serious convulsions, intellectual disability, and other symptoms. Niemann-Pick is a set of serious diseases that includes unsteady gait (ataxia), slurring of speech (dysarthria), difficulty in swallowing, and many other neuromuscular problems.

correct. Sickle cell trait has the effect of being protective against malaria infection, but the price of this protection is a terrible disease. Its existence corresponds not with ethnicity but with the geographical distribution of malaria because they have evolved alongside each other. It is indeed common among people of recent African ancestry, but only those whose descent colocates with malaria zones, which distribution represents as a slice across the middle of the African continent. Similarly, sickle cell disease and trait exist at high frequency in Greece, Turkey, the Middle East, and India, in a pattern that mirrors the range of malaria.

The suggestion by Cochran et al. is that the cost of selection for genes involved in intellectual prowess is a high frequency of a handful of diseases that might be important in brains. As carriers of the disease genes occur at measurable frequencies in the population with little disease effect, it is suggested that these genetic variants are evidence for a genetic basis for enhanced intellect.

They further enhance their argument by suggesting that the supposedly Jewish diseases are ones that are related to a specific type of biochemistry called lysosome storage. Tay-Sachs is a terrible disease that is lethal to children by the time they are three and involves rapid neural degeneration. It was initially identified in Jewish families and did occur at high frequency in Jews, but not exclusively. But that occurrence has been quelled with concerted and careful expert advice about being in high-risk categories—genetic counseling—which has radically reduced the frequency of the disease genes in Jews. Niemann-Pick is typically lethal by eighteen months after birth, and also features neurological degeneration; of the several types of Niemann-Pick, one is most common in Ashkenazim. Gaucher's disease is far less serious, and may have a small neural effect, and is relatively common in—though not exclusive to—Jewish people.

Some other studies showed no signs of selection for the disease genes. Instead, the nature of the mutations in these genes suggested what is known as founder events, that is, new mutations that become fixed in a population, not least because of higher degrees of marrying people within the same family. This is a common occurrence within small and isolated populations. The complexities of the arguments for and against the potential intellectual benefits of these particular genetic conditions are a total mess, in terms of different studies arguing in support of selection, against selection, or for founder effects, genetic bottlenecks, or neutral drift, where changes in DNA are neither beneficial nor detrimental. Cochran et al. suggest other genes or bits of DNA that may be involved in promoting growth of neurons or the dendrites that grow out of them and link to other brain cells.

They did not know this at the time of writing their paper, but we now know that the genes associated with intellectual capability are myriad, and of very small but cumulative effect—pixels on a colossal screen. Of the genes identified so far (and remember that while we know these genes are important, we don't know what they do, and therefore *why* they are important), many are expressed in the brain (as indeed are thousands of genes), and therefore may well have a direct effect on intellect. There are databases that list hundreds of GWAS results and thousands of genes. You can enter a gene and ask the database to pull out studies that indicate the gene is associated with any one of dozens of types of trait, from height to mortality to bones, as well as cognitive and neurological. I checked the current databases for the disease genes that Cochran et al. suggest might be driving selection for Jewish smarts, to see if, at the time of writing, they associated with brains or cognitive abilities. The result? Not one of them does.

Speculation is sometimes an important part of science. Trying to dream up an explanation for an observation can be a productive way of honing a scientific question in the absence of data that explains it. But not in this case. This one paper has a resounding echo, and continues to foster influence and discussion. It was championed by the then science editor of *The New York Times*, Nicholas Wade, in multiple articles and subsequently in a book that was almost universally derided by the genetics community as error strewn and specious, but celebrated by racists. The celebrity psychologist Jordan Peterson uncritically cited Cochran's work in February 2019, when writing about the disproportionate success of Jews in intellectual pursuits.

I don't pretend to know the motivations of people publishing controversial work that does not fare well against the ruthlessness of scientific scrutiny. In my opinion, the Cochran study may appear to be pursuing scientific truth in the face of political correctness; instead it reads as political, but neither true nor scientifically correct.

Mark Twain's sentiment about the disproportionate success of Jews appears true, but if we search for a biological basis for this in cultural domains that require clever brains, the numbers make no more sense. Jews account for a huge number of classical music maestros, soloists, conductors, and players in the very best orchestras. Many of the greatest violinists have been Jewish: Yehudi Menuhin, Itzhak Perlman, Isaac Stern, Jascha Heifetz, and they sit alongside Felix Mendelssohn, Gustav Mahler, Arnold Schoenberg, Leonard Bernstein, András Schiff, Daniel Barenboim, and a host of other performers and composers. Musical talent is more difficult to assess into a single metric than intelligence is, though as ever, there are genetic and environmental factors in musical achievement. But the numbers don't lie: There are effectively no canonical Black classical music composers and few Black members of elite orchestras.*

Classical music is dominated by White people, and Jews are disproportionally successful within this sphere. Yet jazz has been historically dominated by Black musicians. Is there something about jazz that is so inherently different from orchestral music that it must be attended to by biological difference?

No, and sensible people don't make this argument. But there is a common myth that Black people have innate musical abilities—"natural rhythm," so the stereotype goes. This popular assertion of innate talent—meaning encoded in DNA—falls at the slightest hurdle. Jazz, like hip-hop, is a musical genre that emerged, at least in part, as a revolutionary subculture, separate from and in defiance of the prevailing Eurocentric and White American musical styles of the day. Both became enormously popular and went through transitions into mainstream culture, though both were initially feared and prosecuted as dangerous by the authorities at their inception. It would be incredibly easy to show a profoundly strong correlation between leading hip-hop artists and pigmentation

---

*Note though that there are few very well-known or canonical female composers of classical music, nor any women in the list above. Genetically, women are far more different from men than Black men are from White men. Are we to attribute that paucity to the absence of a Y chromosome, or is it more likely because of the fact that women were not allowed to hold this type of position?

genes. Hip-hop remains overwhelmingly dominated by Black artists, though Eminem is quite good. We could do a GWAS on rappers and find correlates with genes associated with being African American. Are we to believe that somehow the genetically encoded aptitude to musical talent extends only to musical styles that Black people excel at, rather than the musical styles in which they are absent? No, because the genetic associations between musical aptitude and ethnicity are unlinked, and the population differences in musical styles can only be a cultural phenomenon.

All human behavior is a heady mix of genes and culture, of biology and history. Not enough is known about genetics or cognitive abilities to make definitive statements about evolutionary selection for genes that enhance the most sophisticated and elegant expressions of humankind. It is possible that there is a genetic nudge in that direction, though unlikely based on current data, and there is very little evidence to support the idea. Instead, to some who profess being only in pursuit of truth, it's an idea appealing enough to warrant endless support. To my mind, commitment to these fingers-crossed speculations says more about the people that hold those views so tenaciously than it does about Jews, Blacks, or any ethnic group. Some of the scientists and race-fixated ideologues are actual racists, others merely contrarians, or skeptics convinced that they have unearthed some secret knowledge that has been quelled by a conspiratorial majority.

Arguments that the presence of brain disorders at high frequency in Ashkenazi Jews might explain the enrichment of genes that boost brains is woolly conjecture, and can be abandoned with current data to hand. "Winnowing through persecution" is also merely idle speculation, and has no place in a decent scientific paper. These are fractionally more sophisticated versions of the evolutionary crime we call adaptationism*—the assumption that natural selection is responsible for specific human behaviors, rather than happenstance or processes that are neither positive nor negative, but have simply drifted into existence. In the genomic age we are capable of actually seeing the parts of the genome where selection has taken place, and there are population-specific mutations that indicate positive selection of particular genes as adaptations to the local environment. Pigmentation, specific diets, resistance to diseases (such as malaria), and other traits are demonstrably local adaptations that are part of humankind's success in colonizing the world. Adaptationism is an error because in many cases it results in untestable hypotheses, but ones that are appealing

---

*Also referred to as "Panglossianism," after Voltaire's character Dr. Pangloss, who thought there is a reason for everything, and thus hypothesized that noses are shaped as they are in order to balance spectacles on them, and that we are bipedal because two legs is what fits a well-cut trouser.

because they sound superficially convincing—Blacks are good sprinters because of selection during slavery; Jews are intellectually gifted because their history of persecution enriched genes associated with brains.

The evidence for selection of genes for intellect in Jews is weak. Is it not simply more scientifically parsimonious to suggest that a culture that values scholarship is more likely to produce scholars? The immense intellectual value placed on the traditions of yeshivot Talmudic scholarship began in the Middle Ages and continues to this day, and is arguably without parallel. Just as in a society that champions long-distance running as a pathway to economic and cultural success, with highly successful runners already in place, a multitude will chase them.

Notionally a positive attribute, an evolutionary history that has fomented intellectual and commercial success is a major part of the standard and long-standing tropes of anti-Semitism. But the argument is riven with anti-Semitic tropes that are ahistorical. Moneylending is a common stereotype, not least because of Shakespeare's Shylock. In fact, moneylending was a trade that was extremely limited in time and space within Jewish culture in Europe, and by the end of the fifteenth century had largely vanished from Jewish populations. Yet the implication of the scientific speculation by Cochran et al. is that business and financial acumen has driven the evolution of Jewish brains.

Wave after wave of grim anti-Semitism is becoming more common in public. Desecration of Jewish graves and swastika graffiti have been reported in the press in 2019 all around Europe. In the US, White supremacists chanted "Jews will not replace us" in Charlottesville in 2017; in 2018 a gunman murdered eleven people and wounded six during Shabbat morning services at the Tree of Life–Or L'Simcha Congregation in Pittsburgh (the deadliest attack on Jews in the country's history). The year 2019 saw at least two anti-Semitic shootings, one inside the Poway synagogue, north of San Diego, and another at a kosher market in Jersey City. Britain is currently politically beset by anti-Semitism, notably centered around the left wing of the Labour Party, one of the two major political parties. Seven members of Parliament resigned from the Labour Party in February 2019, primarily because of the party's failure to deal with rampant anti-Semitism within its ranks. This is the defining issue for Labour in the current era, and in the 2019 general election contributed significantly to their biggest parliamentary defeat in more than eighty years. Anti-Semitism is one of the only forms of racial bigotry that punch upward to *perceived* power, which fuels part of its continued existence within left-wing thinking. Beyond these grotesque politics, the stereotypes of anti-Semitism are based around disproportionate power, wealth, greed, and influence, particularly in the media, in commerce and in politics. An evolutionary basis to Jewish intellectual success serves only to fuel the systematic and historical identification of Jews as separate, different, and powerful. That the disproportionate success of Jews is asserted as innate and evolved into their very existence can be used as a route to differentiating a people as an enemy. That it is portrayed as a trait that comes with a curse of genetic illness is nothing more than a nonscientific fiction.

Genetics, race, intelligence: Marrying these three concepts fails to deliver satisfactory answers. Nor do they overlap in an informative way: Genetic variation in people does not tally with the folk descriptions of race; populations, countries, and continents do vary enormously in average IQ scores, but a genetic explanation struggles to account for the differences; intelligence is heritable, but we have a poor understanding of the genetics that underlies cognitive performance. There are genetic differences between populations, which we can measure, but we do not know what they do. In our studies so far, the nature of these differences reflects different ancestral histories rather than specific phenotypes.

These are not liberal sensibilities; they are merely what the data says, when scientific scrutiny is applied. Science is always provisional, and subject to revision upon the discovery of new facts. Maybe, in the future, patterns will emerge in the ever more precise genomic tools we are inventing, but it seems unlikely in the extreme that our current understanding of the relationship between race, genetics, and intelligence will undergo a radical overhaul.

It might seem odd that a geneticist should want to downplay the significance of genes, but the fact is that we are social beings who have off-loaded so much of our behavior from our bodily hardware to our cultural software, and nowhere is this more apparent than in our intelligence. There is no secret truth waiting to be revealed, no grand conspiracy of silence from geneticists. People are born different, with different innate capabilities and potential. How these abilities cluster within and between populations is not easily explained by fundamental biology, by genetics. Instead, when digging into the data as best as we can, we find the answers not in DNA but in culture.

Conclusion and Recapitulation

We are born with difference coded into our cells. People are born different, look different, and behave differently. We have innate characteristics scored into our DNA. These differences vary between people, and between populations. But, as we have seen, the way we generally speak about races does not align with what we know about those innate differences between people and populations. Genetics and the evolutionary history of humans do not support the traditional or colloquial concepts of race. Here is a recap of the key points.

- **Human variation is real:** Local adaptations in our evolutionary past account for a lot of the physical differences we see in living populations, but certainly not all.

- **The dominance of skin color as a racial classifier is based on historical pseudoscience** primarily invented during the years of European empire building and colonial expansion.
- **The primary physical characteristics of race are not representative of overall similarities or difference between people and populations.**
- **We see broad geographical clustering of people and populations on the basis of sampled genetic markers,** but the borders are fuzzy and continuous.
- **Concepts of racial purity are ahistorical and pseudoscientific.** People move and reproduce with great vigor, and admixture between different and previously separate populations is the norm. That is why humans are so successful.
- **Genetic differences between populations do not account for differences in academic, intellectual, musical, or sporting performance between those populations.**
- **Race is a social construct. This does not mean it is invalid or unimportant.** Humans are social animals, and the way we perceive each other is of paramount importance. But it does mean that the colloquial use of race is a taxonomy that is not supported by our understanding of fundamental biology, meaning genetics and evolution.

I once interviewed a parapsychologist, that is, one who deals in the scientific claims underlying supposedly supernatural phenomena. For many decades, he had conducted ghost-hunting excursions to haunted houses and other eerie spots, in response to claims by a spooked public.

In all that time, he had never seen a single specter. I was left wondering at what point do you call it a day, and just say, "I've looked and I've looked and I can't find any evidence for the existence of ghosts." There are some scientists who seem similarly motivated in the pursuit of a biologically meaningful racial taxonomy. Of course, ghosts don't exist, but people *are* different from each other, and broadly, those overall differences are geographically distributed. As a scientist, one has to remain open-minded, as all results are conditional, and all subject to change as more data becomes available.

People fixated on finding biological bases for racial differences appear more interested in the racism than the science. Arguments in online social media seem to involve people for whom demonstrations of genetic or behavioral differences being evidence for racial categories are the absorbing passion of their lives; these are people who are invigorated by animosity. This is a difficult landscape to navigate, because the vast majority of scientists abandoned the scientific validity of race many years ago, and as a result, very few people in genetics study questions specifically of race. Only the fixated remain, as if they have some secret knowledge that we have been suppressing for reasons of ideology.

This is not my experience. Science is a quintessentially revolutionary process, perpetually trying to overthrow what has come before. But it's also a deeply conservative revolution, meaning that it inches along, chipping away at the current state of knowledge. We're not waiting for a huge revelation about race and genetics, because what is already known accounts for most of what we see in human variation around the world. In physics, the structure of most of the cosmos is unknown, and when dark matter is found in the next few years, our universe will quake. But of the millions of questions that remain about the biological nature of humans, race is not one that is particularly outstanding. Yet because of the social implications of politics, people, history, and power, it remains a defining topic of our age.

Science should in principle be free from prejudice, and we should be led by the data, and not by our political prejudices. I have been accused many times of misrepresenting real science because of political correctness, or because I am mixed race. I have tried to maintain as much scientific integrity as I can muster in a long career in genetics, and to squint at the data, to forensically pick apart what it means. I believe I am honest, and that my motivations are in the service of science.

But nobody's perfect. It is also important to recognize though that while the data should be neutral in principle, it almost never is. As long as humans design experiments, log data, and perform analyses, science is subject to being clotted with prejudice either explicit or unknown to the scientist themselves. The language we use, particularly in genetics, sometimes betrays casual historical racism, which is doubly pernicious. We harvest data from populations that are accessible to us, and fail to consider the boundaries of those samples, yet use a lexicon that reinforces neat but ahistorical clusters: "European," "Caucasian," and other terms that people use to identify themselves may not necessarily represent the ancestries and genetic variance in that group. Yet we continue to embrace them in academic literature, and the result is reinforcement of historical terms, and potentially skewed results. Science should be pure and straightforward, but people are not. We come to these questions with centuries of history behind us, much of which is unknown to most of us but has influenced us in some way.

We celebrate the fact that science is an endeavor built upon past knowledge with the maxim that "we see further by standing on the shoulders of giants." We must also be aware of the centuries of pseudoscience perpetrated by some of the very men of power on whose shoulders we stand, because their ideas have percolated through time, and persist within science and society to this day, no matter how offensive, out of date, or absurd they might be. The sweet irony is that the whole science of human genetics was founded by racists in a time of racism, and singularly has become the field that has demonstrated the scientific falsity of race. As a result, the foundations of racism cannot be drawn from science.

It is the evidence that has led me to think that race is not a biologically useful way of categorizing people. At the same time, I am aware of human differences, in both physical and behavioral traits, and that these differences are interesting. They tell us about our history, our fundamental biology, and our culture. Filtering out nature and nurture is no mean feat. As we delve deeper into the data held in our DNA, we have an increasingly keen sense of the complexity of the relationship between the biological modes of inheritance and the input of the environment in which they play out. But the weight of evidence clearly says that real human variation does not correspond with traditional and colloquial descriptions of race.

Scientists are not naturally predisposed to express political opinions in the wake of their data, as pure or mucky as it might be. I believe in absolute academic freedom of inquiry. I think there are no subjects nor any questions that should be censored from scientific investigation. Studies of biological difference are sometimes controversial, and perhaps some researchers avoid these questions for fear of political backlash. The quality of research is variable, particularly new studies of cognitive abilities enabled by the vast repositories of DNA now available for all to plunder. Sometimes these studies are statistically underpowered, or draw conclusions from sample sizes that cannot support those conclusions. Sometimes the wrong tools are deployed in addressing a scientific question. It falls to other scientists, often busy doing their own unrelated research, to criticize lousy science, while simultaneously the Internet has enabled the rapid and open dissemination of content of all levels of integrity and quality. These are new terrains for us to negotiate.

But scientists are only a cog in the framework of structural racism that permeates our society. The small number of fringe researchers who continue to pursue a biological basis of race and the marginalized extremists in the form of White nationalists are foes worth confronting, because their voices can have the effect of normalizing racist attitudes among the wider public. Prejudices are a natural part of the human condition, and much of the science that undermines the biology of race may run counter to your experience, which sits alongside all those historical biases that are baked into our culture.

The misunderstandings of genetics and of genealogy are a fuel for reinforcing racism. This is a problem unique to the biological sciences. If a physicist were to tell you about the behavior of quarks and how that relates to the fundamental structure of matter, you would have to take a lot of it on trust, because the complexities of the quantum realm are impenetrable to all but a few specialists, and bear little relationship to your personal experience. Furthermore, why would you have any reason to doubt? There's no political motivation or prejudice in explaining the nature of matter or space-time.

But human genetics is us. Our senses, our experience, and our culture steer us in our thinking about fellow humans. There may well be evolutionary and psychological reasons for these prejudices, such as to protect ourselves and kin from invaders who want to kill us and take our stuff. And there are clear political motivations underlying discussions of human variation and race. You might be a racist, whose intentions are to persecute or quell the potential success of people with ancestral trees different from your own. You might be a liberal who wants to promote equality, and is therefore also content with misrepresenting what can be known.

Or maybe you are just simply casually confirming your own biases, and in doing so affect racist views. It is easier to fixate on one particular phenotype, such as skin color, or one individual gene, such as *ACTN3*, or one metric, such as IQ, and hang all your prejudices—conscious or otherwise—on them, rather than scrutinize the deeper reality. It is easier to apply "common sense" arguments, such as that slavery bred natural athletes, than recognize that life histories, evolution, and genetics are profoundly tricky to unpick. And it is certainly damnably easier to use new genetic techniques to see patterns in data that superficially reinforce stereotypes rather than apply fiendishly difficult statistics to show that they are not meaningful. All of these wretched traps are rooted in science, and demonstrate bad scientific practice, but they are views commonly held by people who are fixated on finding biological evidence to support their ideas about race.

The observations of difference in whatever metric—sport, intelligence, musical ability, skin color—represent the *beginning* of scientific inquiry, not an endpoint. Good science is wont to disentangle our observations, to establish how things really are, rather than how we perceive them to be, and to lean in toward the truth rather than assert it. This is why the terms "scientific racism" and "race science" are both misnomers. These are pseudoscientific domains. The lessons of contemporary genetics are essential in preventing pseudoscience from seeping deeper into the cracks in our societies, and wedging more division into our lives.

It behooves us all to confront racism wherever we find it, especially when it is covert or normalized in stereotypes and myth, and science is a weapon in that contest. The academic and political activist Angela Davis said that "in a racist society it is not enough to be non-racist, we must be anti-racist."

Of course, racism is not wrong simply because it is based on scientifically specious ideas. Racism is wrong because it is an affront to human dignity. The rights of people and the respect that individuals are due by dint of being a person are not predicated on biology. They are human rights. Hypothetically, if there were genetic differences between populations that we have not found yet, and these do correspond with the folk definitions of race, the fact that we have not found them means they are tiny at best. If those things were true—and there is no evidence that they are—would that have any impact on how we should treat each other? If science were somehow to show that there are genetic differences that align with our folk use of the terms of race, and that these also account for perceived differences in ability, would that justify segregation? Would you afford people different rights if they were ancestrally faster, brighter, or stronger?

Imagined differences between individuals and between populations have been used to justify the cruelest acts in our short history. Learned prejudices fuel bigotry, which will inevitably continue. What is important for science is that we recognize and study the reality of biological diversity in order to understand it, and consequently to undermine its bastardization.

Race is real because we perceive it. Racism is real because we enact it. Neither race nor racism has foundations in science. It is our duty to contest the warping of scientific research, especially if it is being used to justify prejudice. If you are a racist, then you are asking for a fight. But science is my ally, not yours, and your fight is not just with me, but with reality.

Acknowledgments

The following people have helped me formulate my ideas either directly by peer-reviewing earlier drafts, or simply through their work, and I am grateful to each of them. The work is theirs; any errors are mine: Andrew Chukwuemeka, Caroline Criado Perez, Caroline Dodds Pennock, David Epstein, Simon Fisher, Hannah Fry, Alex Garland, Ben Garrod, Natalie Haynes, Nina Jablonksi, Steve Jeal and the Elite U12 Old Alleynians coaching squad, Greg Jenner, Steve Jones, Debbie Kennett, Tracy King, David Lammy, Alex Lathbridge, Tina Lasisi, Nathan Lents, Helen Lewis, Ana Paula Lloyd, William Mathew, Aoife McLysaght, Elspeth Merry-Price, Kevin Mitchell, David Olusoga, Aaron Panofsky, Robert Plomin, Danielle Posthuma, David Reich, Stuart Ritchie, Ananda Rutherford, David Rutherford, Aylwyn Scally, Francesca Stavrakopoulou, Adrian Timpson, Cathryn Townsend, Tupac Shakur, Lucy van Dorp, Tim Whitmarsh, and

Alun Williams. I am particularly indebted to Jennifer Raff, Ewan Birney, Graham Coop, Mark Thomas, Jedidiah Carlson, Alice Roberts, and Matthew Cobb, whose kind scrutiny is as sharp as a scalpel. As ever, the people who fundamentally enable me to write, and make my life joyous, all have the same surname as I do: Georgia, Beatrice, Jake, and Juno.

Though it is my name alone on the cover, this book is the product of teamwork from dedicated editors, designers, agents, and friends. My editorial team is a joy to work with; via careful thought, argument, discussion, and editorial panel beating, they do nothing but make what I do considerably better: PJ Mark and Will Francis at Janklow & Nesbit, in the UK, Jo Whitford, Holly Harley, and Jenny Lord, and for The Experiment, Matthew Lore, Jennifer Hergenroeder, and above all, Nick Cizek for his guidance and craft and for pointing out eighty-seven times that football and soccer are not the same thing.

References

Andrews, Kehinde. “From the ‘Bad Nigger’ to the ‘Good Nigga’: An Unintended Legacy of the Black Power Movement.” *Race and Class* 55, no. 3 (October 24, 2013): 22–37.

Ash, G. I. et al. “No Association between ACE Gene Variation and Endurance Athlete Status in Ethiopians.” *Medicine and Science in Sports and Exercise* 43, no. 4 (April 2011): 590–7.

Bhatia, Gaurav et al. “Genome-Wide Scan of 29,141 African Americans Finds No Evidence of Directional Selection since Admixture.” *American Journal of Human Genetics* 95, no. 4 (October 2014): 437–44.

Cochran, Gregory et al. “Natural History of Ashkenazi Intelligence.” *Journal of Biosocial Science* 38, no. 5 (September 2006): 659–93.

Crawford, Nicholas G. et al. “Loci Associated with Skin Pigmentation Identified in African Populations.” *Science* 358, no. 6365 (November 17, 2017).

Dover, Cedric. “The Racial Philosophy of Johann Herder.” *British Journal of Sociology* 3, no. 2 (June 1952): 124–33.

Edge, Michael D. and Graham Coop. “Reconstructing the History of Polygenic Scores Using Coalescent Trees.” *Genetics* 211, no. 1 (January 2019): 235–62.

Fan, Shaohua et al. “Going Global by Adapting Local: A Review of Recent Human Adaptation.” *Science* 354, no. 6308 (October 7, 2016): 54–9.

Gayagay, G. et al. "Elite Endurance Athletes and the ACE I Allele—The Role of Genes in Athletic Performance." *Human Genetics* 103, no. 1 (July 1998): 48–50.

Helgason, Agnar et al. "A Populationwide Coalescent Analysis of Icelandic Matrilineal and Patrilineal Genealogies: Evidence for a Faster Evolutionary Rate of mtDNA Lineages Than Y Chromosomes." *American Journal of Human Genetics* 72, no. 6 (June 2003): 1370–88.

Hughey, Matthew W. and Devon R. Goss. "A Level Playing Field? Media Constructions of Athletics, Genetics, and Race." *The ANNALS of the American Academy of Political and Social Science* 661, no. 1 (August 10, 2015): 182–211.

Kuncel, Nathan R. and Sarah A. Hezlett. "Fact and Fiction in Cognitive Ability Testing for Admissions and Hiring Decisions." *Current Directions in Psychological Science* 19, no. 6 (December 14, 2010): 339–45.

Lapchick, Richard with Angelica Guiao. "The 2015 Racial and Gender Report Card: National Basketball Association." The Institute for Diversity and Ethics in Sport (TIDES) (July 1, 2015): 1–44.

Ma, Fang et al. "The Association of Sport Performance with ACE and ACTN3 Genetic Polymorphisms: A Systematic Review and Meta-Analysis." *PLoS ONE* 8, no. 1 (January 24, 2013).

Martin, Alicia R. et al. "Human Demographic History Impacts Genetic Risk Prediction across Diverse Populations." *American Journal of Human Genetics* 100, no. 4 (April 6, 2017): 635–49.

Martin, Alicia R. et al. "An Unexpectedly Complex Architecture for Skin Pigmentation in Africans." Cell 171, no. 6 (November 30, 2017): 1340–53.

Panofsky, Aaron and Joan Donovan. "Genetic Ancestry Testing among White Nationalists: From Identity Repair to Citizen Science." *Social Studies of Science* 49, no. 5 (July 2, 2019): 653–81.

Patin, Etienne et al. "Dispersals and Genetic Adaptation of Bantu-Speaking Populations in Africa and North America." *Science* 356, no. 6337 (May 2017): 543–6.

Peterson, Jordan B. "On the So-Called 'Jewish Question.'" Jordan B. Peterson (blog), February 26, 2019. https://jordanbpeterson.com/psychology/on-the-so-called-jewish-question/.

Pickering, Craig and John Kiely. "ACTN3: More Than Just a Gene for Speed." *Frontiers in Physiology* 8, no. 1080 (December 18, 2017).

Pickrell, Joseph K. and David Reich. "Toward a New History and Geography of Human Genes Informed by Ancient DNA." *Trends in Genetics* 30, no. 9 (September 2014): 377–89.

Polderman, Tinca J. C. et al. "Meta-Analysis of the Heritability of Human Traits Based on Fifty Years of Twin Studies." *Nature Genetics* 47, no. 7 (May 2015): 702–9.

Redfern, Rebecca et al. "A Novel Investigation into Migrant and Local Health-Statuses in the Past: A Case Study from Roman Britain." *Bioarchaeology* International 2, no. 1 (July 2, 2018): 20–43.

Rosenberg, Noah A. et al. "Genetic Structure of Human Populations." *Science* 298, no. 5602 (December 20, 2002): 2381–5.

Rosenberg, Noah A. et al. "Interpreting Polygenic Scores, Polygenic Adaptation, and Human Phenotypic Differences." *Evolution, Medicine, and Public Health* 2019, no. 1 (December 2018): 26–34.

Sagnella, Giuseppe A. et al. "A Population Study of Ethnic Variations in the Angiotensin-Converting Enzyme I/D Polymorphism: Relationships with Gender, Hypertension and Impaired Glucose Metabolism." *Journal of Hypertension* 17, no. 5 (May 1999): 657–64.

Scott, R. A. et al. "No Association between Angiotensin Converting Enzyme (ACE) Gene Variation and Endurance Athlete Status in Kenyans." *Comparative Biochemistry and Physiology: Part A, Molecular and Integrative Physiology* 141, no. 2 (June 2005): 169–75.

Scott, R. A. et al. "ACTN3 and ACE Genotypes in Elite Jamaican and US Sprinters." *Medicine and Science in Sports and Exercise* 42, no. 1 (January 2010): 107–12.

Ségurel, Laure et al. "The ABO Blood Group Is a Trans-Species Polymorphism in Primates." *Proceedings of the National Academy of Sciences* 109, no. 45 (November 6, 2012): 18493–8.

Skoglund, Pontus and David Reich. "A Genomic View of the Peopling of the Americas." *Current Opinion in Genetics & Development* 41 (December 2016): 27–35.

Sniekers, Suzanne et al. "Genome-Wide Association Meta-Analysis of 78,308 Individuals Identifies New Loci and Genes Influencing Human Intelligence." *Nature Genetics* 49, no. 7 (July 2017): 1107–12.

Tishkoff, Sarah A. and Kenneth K. Kidd. "Implications of Biogeography of Human Populations for 'Race' and Medicine." *Nature Genetics* 36 (2004): S21–S27.

Trahan, Lisa et al. "The Flynn Effect: A Meta-Analysis." *Psychological Bulletin* 140, no. 5 (September 2014): 1332–60.

Van Dorp, Lucy et al. "Evidence for a Common Origin of Blacksmiths and Cultivators in the Ethiopian Ari within the Last 4500 Years: Lessons for Clustering-Based Inference." *PLoS Genetics* 11, no. 8 (August 20, 2015).

Van Dorp, Lucy et al. "Genetic Legacy of State Centralization in the Kuba Kingdom of the Democratic Republic of the Congo." *Proceedings of the National Academy of Sciences* 116, no. 2 (January 8, 2019): 593–8.

Veale, David et al. "Am I Normal? A Systematic Review and Construction of Nomograms for Flaccid and Erect Penis Length and Circumference in Up to 15,521 Men." *BJU International* 115, no. 6 (December 2014): 978–86.

Webborn, Nick et al. "Direct-to-Consumer Genetic Testing for Predicting Sports Performance and Talent Identification: Consensus Statement." *British Journal of Sports Medicine* 49, no. 23 (December 2015): 1486–91.

Wicherts, Jelte M. et al. "A Systematic Literature Review of the Average IQ of Sub-Saharan Africans." *Intelligence* 38, no. 1 (January–February 2010): 1–20.

Wilber, R. L. and Y. P. Pitsiladis. "Kenyan and Ethiopian Distance Runners: What Makes Them So Good?" *International Journal of Sports Physiology and Performance* 7, no. 2 (June 2012): 92–102.

Witherspoon, D. J. et al. "Genetic Similarities within and between Human Populations." *Genetics* 176, no. 1 (May 2007): 351–9.

Image Credits

Page 54
Distributed under the CC BY 4.0 license.

Page 55
Courtesy of the Bibliothèque Nationale de France.
Public domain.

Page 73
Photograph copyright © Kennis & Kennis Reconstructions

Page 94
Howard Davies/CORBIS/Corbis via Getty Images

Page 127
Bettmann/Getty Images

Page 128
Public domain

Page 130
PCN Photography/Alamy Stock Photo

Page 142
Guo Chen/Xinhua/Alamy Live News

Index

NOTE: Page numbers in *italics* indicate an image. Page numbers followed by "n" indicate a footnote.

C

D

G

H

I

J

K

L

P

R

T

U

V

W

Y

About the Author

Dr. Adam Rutherford is a geneticist, science writer, and broadcaster. He studied genetics at University College London, and during his PhD on the developing eye, he was part of a team that identified the first known genetic cause of a form of childhood blindness. As well as writing for the science pages of the *Guardian*, he has written and presented many award-winning series and programs for the BBC, including the flagship weekly Radio 4 program *Inside Science*, *The Cell* for BBC Four, and *Playing God* (on the rise of synthetic biology) for the leading science series *Horizon*. He is the author of *The Book of Humans*, a new evolutionary history that explores the profound paradox of the "human animal"; *A Brief History of Everyone Who Ever Lived*, finalist for the National Book Critics Circle Award in nonfiction; and *Creation*, on the origin of life and synthetic biology, which was short-listed for the Wellcome Book Prize.

@AdamRutherford